TRIVENT

PUBLISHING

TRIVENT
Ethics | in Science & Technology

Ethics of Emerging Biotechnologies

From Educating the Young to Engineering Posthumans

Edited by

Maria Sinaci & Stefan Lorenz Sorgner

Applied Ethics: From Bioethics to Environmental Ethics, vol. 1

Ethics of Emerging Biotechnologies:
From Educating the Young to Engineering Posthumans
Ed. Maria Sinaci, Stefan Lorenz Sorgner
ISBN 978-615-6405-04-3
Applied Ethics: From Bioethics to Environmental Ethics
Series ed. János I. Tóth, responsible publisher: T. Artimon
ISSN 2630-8185
https://trivent-publishing.eu/45-applied-ethics

Available online at http://trivent-publishing.eu
Cover: CC0

Trivent Publishing
1119 Budapest, Etele út 59–61.
Hungary

TABLE OF CONTENTS

Part IV
End of Life Ethical Issues

Part V
Bioethics and Regional Education

Part VI
Legal, Political, and Social Bioethical Issues

INTRODUCTION

Pastor Fritz Jahr used the term bioethics as early as 1927 when he published an ethical study on human relationships with animals and plants: *Bio-Ethik. Eine Umschau über die ethischen Beziehungen des Menschen zu Tier und Pflanze.*[1] However, it was not until the early 1970s that the term "bioethics" was rediscovered in the United States. Since then, the academic discipline of bioethics has permanently gained more international relevance due to ground-breaking scientific discoveries, medical innovations, and biotechnological inventions. The book *Bioethics. Bridge To The Future* by R. van Potter was of central relevance in the development of bioethics as a separate academic discipline, which is less than 50 years old. Herein, the need of "a new field to combine biological knowledge with the knowledge of the human value system" was stressed·.[2] In this context, bioethics took shape as a new research area striving to give answers, from an ethical perspective, to new environmental and biomedical challenges, such as human reproduction technologies, organ transplantation, stem cell therapy, and so on. Furthermore, bioethics discusses issues of human genetics, genetic information ownership and usage, genetic modifications, patient confidentiality, human enhancement. An essential characteristic of bioethics is its interdisciplinary approach, including distinctive fields such as medicine, law, sociology, philosophy, psychology, and theology. Although mostly materialized in Western Europe and North America, the field of bioethics now has a global and highly-topical dimension which stresses the need of knowledge on new ethical issues produced by the progress in medicine, bioengineering, pharmacology, genetics, but also by changes in society.

This volume is composed of selected papers from the First International

---

[1] Fritz Jahr, "Bio-Ethik. Eine Umschau über die ethischen Beziehungen des Menschen zu Tier und Pflanze" *Kosmos. Handweiser für Naturfreunde* 24(1) (1927): 2–4.

[2] Van Rensselaer Potter, *Bioethics: Bridge to the Future* (Englewood Cliffs: Prentice Hall, 1971).

Conference on Bioethics in the New Age of Science (BNAS), which took place on the 4th and 5th of May 2017 at the "Vasile Goldis" West University of Arad, Romania. The papers were extended into chapter-length contributions. Furthermore, the volume includes contributions from additional scholars with expertise in the multi-faceted field of bioethics. The volume is divided in six parts which go through subjects ranging from enhancement technologies, to medical and legal issues in bioethics, as well as aspects concerning bioethics education.

The first part is an introduction on the ethical challenges of emerging (bio)technologies and the authors of the three chapters leap into discussions on gene ethics, CRISPR/Cas9, posthumans (Stefan Lorenz Sorgner summarizes the most relevant ethical challenges of these issues, both present and future), on philosophical issues of techno-science (where Corrado Viafora discusses the topical issue of the ethical capacity to govern technology), and the moral relationship between human and non-human animals (where Alexander Krémer argues through a Kantian perspective that although animals cannot be moral agents, humans must be moral towards them in order to maintain human dignity).

The second part is a continuation of the first one as its two chapters put into perspective specific aspects of enhancement technologies. Maria Sinaci thus brings forwards a thought-provoking subject, that of moral enhancement through direct interventions, in her chapter titled *Neuroethics and Moral Enhancement: The Path to Moral World?*, while Ivan Mladenović discusses the role of public deliberation on bioethical issues – specifically, the author refers to public deliberation with regard to human biomedical improvements.

The third part dives into specific cases of medical bioethics. G. Tari and G. Braunitzer highlight the conflict between the principles of *beneficence* and *patient autonomy*, referring to bilateral mastectomy in their chapter titled *On the ethical issues of bilateral and contralateral risk-reducing mastectomy*. The second contribution to this part focuses on patient quality of life hematological patients – the research of Alciona Sasu et al. concluded that the most important factors which can change the quality of life in an onco-hematological patient are biological, psychological, and social.

Part four, related in certain aspects to part three, debates on end of life issues. Assya Pascalev analyses here the role of metaphors in biomedical ethics, using as case study the dominant metaphors of death and dying employed by critical care physicians, while Florin Lobont and Ionut Mladin brings into attention the importance of philosophical and ethical counselling for both terminally-ill patients and medical personnel.

The penultimate part focuses on bioethics education with two regional case studies. Thus Antoanela Naaji presents her research study on the perception of the usage and relevance of bioethics courses in specialised academic programmes in Western Romania; while Dejan Donev outlines the introduction and evolution process of (bio)ethical education in Macedonia, starting with 1998, as well as the awareness and interest it generates among young people.

The last part of the volume is a mixture of legal, political, and social bioethical issues. Miroslav Radenković et al. approach the problematic of mandatory vaccination of healthcare personnel – an action which, as the authors argue, can be justified from an ethical point of view when the harm to patients and the overall population is considered to outweigh the autonomy of the individual healthcare practitioner. Further on, Jakub Berezowski discusses the regulating process of the Polish medical legislation. The author highlights the importance of the medical procedure application growth as this would increase the safety of both patients and medical personnel. In *A New Pro-Natalist Tool: Parenthood as a Form of Public Employment*, János I. Tóth marks the problem of low demographics and fertility rates in Europe and Eastern Asia. The author proposes a possible radical pro-natalist institution which would consider parenthood as form of public employment. The last chapter is authored by Maria Ancuța Gurza, *Cultural and Ethical aspects of social desirability in psychological research*, and investigates the child-parent relationship from the perspective of the attachment theory and parenting style, with reference to both foster and biological parents.

The relevance of bioethics as an academic discipline can hardly be underestimated. The age of biotechnological and medical innovations has only just begun, and enormous progress can be expected in various areas relevant to bioethical discourses in the coming decades and centuries, which is why the intensity of the discussions is likely to increase. In particular, questions that fall under the term enhancement will probably become significantly more relevant, such as: 1. Genetic enhancement (CRISPR/Cas9; 2. Pharmacological enhancement (Ritalin, Modafinil); 3. Morphological enhancement (plastic surgery); 4. Cyborg enhancement (brain-computer interfaces) where Cyborgs are cybernetic organisms, i. e. complexes consisting of living organisms and digital or non-digital machines. Increased further technological developments could pose new and fundamental challenges that may call into question our self-understanding as human beings. If cognitive abilities were to be significantly improved, if the human health-span widened and if new emotional states were to be developed, could we still speak of human beings, or would this be linked to the emergence of the transhuman? What ethical challenges would there be, if it were possible to create individuals of a new species, i. e. posthumans, with the help of biotechnologies? Is there a danger of establishing new, self-contained classes, as described in the novel *Brave New World* or the film *Gattaca*? Are we obliged to attribute post-personhood to trans- and posthumans? These perspectives are of enormous urgency and relevance, since they go beyond the narrow realm of bioethics and confront people with fundamental philosophical, biological, anthropological, theological, and political challenges. These and other topics which are related to emerging technologies will keep us occupied for many years to come.

Maria Sinaci & Stefan Lorenz Sorgner

PART I

GENERAL ETHICAL CHALLENGES OF EMERGING (BIO)TECHNOLOGIES

Genes, CRISPR/Cas 9, and Posthumans

Stefan Lorenz Sorgner

We live in an age where paradigm shifts occur in numerous fields of the life world. Most recently, a news report went through the media that a biobag was invented in which a little sheep was successfully grown. This invention has potentially enormous implications, in particular, if we combine this technology with further advances in the field of gene technologies. Discussing the exponential growth of emerging technologies and correlated changes does not imply a derogatory judgement concerning these developments. On the contrary, artificial wombs might be the key technology concerning gender equality at the workplace. If combined with further gene technologies, it has the potential to actively imply radical consequences concerning the future of humanity, an option we have never had before. Nevertheless, these promises as well as the challenges related to enhancing evolution can be expected soon, and the coming about of the posthuman perspective is particularly relevant in this context. It is related to the following insights:

1. A move from a dualist anthropology towards a non-dualist one
2. A radical increase of anthropotechniques with the power of enhancing human capacities so that the likelihood of the posthuman coming about can be increased
3. The posthuman can come about by means of digital technologies or with the help of biotechnologies whereby the field of genetics, on which I focus here, is particularly relevant

These three basic insights lead to a massive number of intellectual, social, political, ethical, and economic challenges [1]. In the following reflections,I will present a selection of emerging bioethical issues concerning gene technologies, in particular CRISPR/Cas 9, and how they can be addressed in an appropriate manner within a wider cultural context. Taking the stance that these developments should not occur and demanding that these technologies should not get developed is not a realistic option. If these technologies are not developed in one country, they will be realized in another one. If scientists and engineers do not receive permission to conduct research in any country, they will create permanent dwellings at sea outside the realms of national governments, i.e. by means of sea-steading.

Non-dualist understanding of human beings

Zarathustra might have been the first to create a rigid distinction between good and evil. This categorical distinction in most philosophies goes along with categorical dualist ontological distinction, and in particular with the understanding that all human beings consist of an immaterial soul, consciousness or mind, and a material body. This insight has been dominant in Western philosophies since at least Plato, who clearly separated the realm of forms which can be accessed by means of thinking from the sensual realm in which we live. This basic insight was accepted by most Western philosophers. Each one adapted it to their own individual understanding. According to both Plato and the Stoics, all human beings have a rational soul. However, for the Stoics, this is the reason why all human beings should be considered moral equals, too, which was not the case according to Plato's thinking. Descartes agrees with the stoic insight of human beings possessing a rational soul. However, he holds that only human beings have any type of soul. Animals consist of matter only. The Stoics and Plato, on the other hand, claim that there are further types of souls which animals and plants possess. In many respects, Kant agrees with Descartes' understanding of animals and human beings, but Kant uses these insights for developing a complex ethics on this basis, an ethics which is still widely taken for granted today [2]. Even the German foundational law has its intellectual basis on the concept of human dignity, a concept which can only be fully understood when having a grasp of Kant's ethics [3]. Only human beings have dignity. Animals, plants, and stones should be treated like things. They all fall under the object law. It is the distinction between subjects and objects or, in moral terms, between things and persons – an idea which has its intellectual root in the dualist anthropological tradition which has been dominant in Western countries since at least Plato. It also has significant practical implications raging from the prohibition of peep shows to the prohibition to shooting down hijacked planes which fly into nuclear power stations to the moral status of animals and how animals can be treated within experiments.

However, this anthropology has become less and less plausible from the end of the 19th century onwards due to insights put forward by Darwin, Nietzsche, and Freud. We no longer regard ourselves as being categorically separate from this world concerning our ontological status – we no longer hold that we have a material body and an immaterial soul. It is a more modest way of thinking, as it moves away from the traditional human hybris of possessing a categorically special status in the world. This does not mean that we do not possess special qualities. Learning a human language might merely be possible for human beings. However, animals also have such special qualities. Vampire bats manage to detect blood by means of infrared sensors. It is their special capacity. Animals, too, can have special abilities. Moving away from the traditional anthropology implies that it is no longer plausible that only human beings participate in the immaterial realm, and are consequently attributed personhood, whereas all other beings are seen as things or objects. However, this is still the qualification which is legally valid in Germany and in many other countries.

I am not claiming that we should replace a dualist anthropology with a non-dualist one concerning the legal realm, because in this way we would replace one fundamentalist view with another one, whereby neither one is shared by all citizens. However, this insight leads to my suggestion that such strong anthropologies and ontological positions should not be part of liberal democracies, because they come in conflict with the plurality of world views which we can find in all liberal democracies today. This shift is fundamentally relevant for all the other developments I will be presenting. Seeing ourselves as merely gradually separate from all other animals implies that, in the same way as they have come about on the basis of evolutionary processes, this also applies to us. In the same way as all other species of animals can die out, this insight also applies to us, if we do not adapt appropriately. If we do not adapt to the permanently-changing environment, we will die out. However, if we do adapt,, we will develop further, via transhumans, i.e. further developed humans, and, if we are lucky, towards the posthuman, i.e. becoming members of a new species. This is one reason why technologies are of immense relevance, but it is by far not the only reason for using technologies [1].

Personhood for animals, robots, and artificial intelligence (AI)

One of the implications of the revised understanding of human beings, as described above, is the relevance of moving away from the anthropocentric basis for evaluating the moral status of an entity, as it implies speciesism, which has morally-problematic implications. Peter Singer was right when he explained that attributing personhood solely and exclusively to human beings implies speciesism. Moral recognition should depend on morally-relevant capacities and not solely on someone's belonging to a specific species. I am not embracing Singer's counter suggestion, which focusses on the relevance of self-consciousness and sentience, as these imply an intense capacity of experiencing pain, and this ability should be the basis of a moral status [5] [6]. Still, it is a strong suggestion. Self-consciousness is a fascinating trait. Whoever possesses it seems to be more prone to suffering than those who do not have it [7] (Morton 2000). Furthermore, the capacity to suffer seems to be relevant for evaluating the moral status of an entity. The mirror test is widely taken for granted as a reliable way for testing whether an organism has self-consciousness. "Currently, 9 non-human animal species pass the mirror test. Not all individuals of each species pass, but many do" [8]. Still, one could wonder whether the test is biased on the eyes. Maybe, self-consciousness in some organisms is more closely related to hearing or smelling, e.g. dogs who do not pass the mirror test. What alternative tests could be applied? Furthermore, there seems to be a close connection between the mirror test, self-consciousness, and our current selfie culture. Speaking of this, can Dürer's self-portrait be analysed as a selfie?

Yet, Singer's suggestion has implications which are not regarded as plausible by most enlightened people. By means of his theoretical stance, it is possible to morally justify the killing of a new-born human being which seems counter-intuitive. However, our widely shared emotional reaction towards this consequence might just be our lack of emotional morality, as we have been overtly and strongly shaped by the traditional Judeo-Christian tradition. If this is the case, it might just be a matter of time until we will have adapted our emotional morality to our theoretical morality.

In any case, Singer's suggestion represents a move into the right direction. An interplay between three pillars (widely shared moral intuitions, the latest scientific insights, and a recognition of the relevance of negative freedom) should provide us with a basis for evaluating the moral status of any type of entity [9]. In addition, cultural embeddedness needs to be taken into the consideration. My approach is narrative and hermeneutic, stressing the relevance of discourse plus the recognition of the relevance of the aforementioned three pillars. This position takes into consideration that social situations and moral evaluations are permanently subject to change and that such changes are relevant for the legal evaluation of entities. In addition, it is a procedural solution which does not aim for a perfect state and solution because it recognizes the relevance of movement and change in the field of morality.

This position can also integrate new developments, like the option of creating hybrids. The UK has permitted to create chimeras consisting of animals and humans, if they are destroyed within two weeks after their realization. However, why should they have to be destroyed? We lack a basis for evaluating the moral status of these hybrid entities. They do not go against human dignity, as these entities are no human beings. The potential of these beings can be enormous. Dutch scientists have already managed to genetically engineer zebra fish so that they can use photosynthesis for nourishment purposes. The fish turn slightly green as part of this process, but it works. Genetically zebra fish are not so different from human beings. Maybe, the little green human beings from Mars are actually our future [2, p. 169-171].

Furthermore, it must be considered that we are already hybrids. Our skin and in intestines contain an enormous amount of bacteria and other microbes. A human body consists of more non-human cells than of human cells, and we could not survive without these cells. This insight can be particularly relevant for the future of xenotransplantation. Martine Rothblatt, a transgender transhumanist, owns a pig farm for genetically engineering pigs so that their lungs can be transplanted into human beings without the risk of them being rejected. It seems like a promising attempt. She is particularly interested in this procedure as her daughter is suffering from a life-threatening lung disease.

However, moving away from attributing a categorically-special status solely to human beings can also imply that, given the appropriate developments, we may have to attribute personhood to computers or AI. Researchers are already trying to see whether consciousness is a phenomenon based on the complexity of neuronal structures by attempting to imitate the complexity of a cat's brain using one computer per neuron. However, consciousness may not even be needed gaining a special moral status. How should we treat a computer with super-intelligence?

Katherine Hayles provides reasons for claiming that there can be something like non-conscious cognition, and if this is the case, it might be plausible to expand the circle of entities which should be morally considered [10]. This thought is based on the following reflections. If we are humiliated, we do not necessarily experience the feeling of suffering. Being humiliated might simply be connected to the cognitive realisation that one is not valued by someone else, and we do not appreciate not being valued. Humiliation could be a problematic relation not only for self-conscious beings, but also for non-conscious, but cognitive entities, i.e. androids like Data from Star Trek. If this is the case then it could make sense to claim that the cognitive capacity to grasp how one is valued is another relevant basis for evaluating the moral status of an entity, which could be a basis for attributing a moral status to certain robots. In any case, it needs to be noted that the move away from an anthropocentric world-view and the corresponding new understanding of a moral status is central for future bioethical challenges, and it is of particular relevance when we deal with moral issues concerning gene technologies.

The dissolution of the moral prohibition to treat a person as merely a thing, as both personhood as well as thinghood no longer exist in their traditional form

The person-thing distinction, which goes back to Kantian philosophy, has numerous general moral implications [3]. One of them is the moral prohibition to treat a person as merely a thing. In Germany, this principle implies that peep shows are legally forbidden and that it is forbidden to shoot down a hijacked plane which seems to be directed towards a nuclear power station as long as there is at least one innocent being on board. In both cases, a person would be treated as merely an object. In the case of the hijacked plane, the innocent pilot, for instance, would die anyway but millions of lives could be saved, if the plane was shot down. However, this would imply that the government treated the innocent pilot as an object in order to save the lives of a million people. Such utilitarian calculations go against the ethics of human dignity and the traditional understanding of personhood which leads to the above-mentioned moral prohibition. However, if the traditional person-thinghood distinction is no longer plausible, the moral prohibition is no longer applicable either. Hence, we need to find a new basis for moral principles which is not an easy task, because the traditional understandings of personhood and thinghood are part of many legal constitutions and have highly significant implications for our life world. It is a task with which we need to deal with in order to morally shape the future. Do we treat a fertilized egg solely as a means, when we select it after in vitro fertilization and preimplantation genetic diagnosis? Do parents treat their offspring just like an object, if they decided to genetically modify them? [11].

The plurality of the good (cognitive capacities, morality, and new family concepts)

Once we take this new anthropology seriously and we understand that all aspects of our existence participate in evolutionary processes, the following question needs to be asked anew: What can be said about living a good life? Some transhumanists suggest strong concepts of the good, like the validity of the Renaissance ideal. It might not be possible for all of us, but we do desire to be intelligent, beautiful, strong, healthy, and have all the other strengths which are associated with the Renaissance ideal of human perfection. Other liberal or naturalist bioethicists, such as Julian Savulescu [12] [13], are more sympathetic with a common sense approach. You live a good life, if you possess the following qualities:

1. Not being disabled whereby disability is seen as context-dependent quality;

2. No disposition for mental illnesses;
3. Having good health;
4. Good capacities for communication, memory, and empathy;
5. High intelligence.

I do not regard these suggestions as plausible, as universal validity implies their application to all human beings at all times in all parts of the world. I, on the other hand, regard a radically-pluralist account of the good life as most plausible [14]. Only by listening to and acting in accord with someone's psychophysiological demands does a person become authentic. Consequently, the following acts can be understood as based on authentic wishes (this does not need to be the case in all instances, but can be): (1) Person A wishing to die; (2) Person B desiring to have her healthy leg removed; (3) Person C wishing to eat parts of himself; (4) Person D not wanting to be cured from her manic depression; (5) Person E regarding his deafness as an advantage and not as a disablement. The list of potential examples could easily be continued. If the Renaissance genius account or the commonsense account of the good were universally valid, these wishes could not be accepted as authentic, but would be seen as expressions of an ill mind.

I do not think that this has to be the case. By claiming that these wishes represent insane states of the mind, these persons are treated paternalistically and violently: their wishes are not recognized as their own, and others claim that they know better than oneself what is in one's own interest. I regard such a way of treating people as highly problematic because the otherness of someone else's wishes does not get appropriately appreciated. On the one hand, there are culturally-dominant paradigms of leading a good life; on the other hand, there are the needs of one's own psychophysiology that do not necessarily correspond to these general demands. A pregnant woman who wishes to have sex with men other than the father of her child, a student who enjoys sexual intercourse with several people at the same time, and a young girl who is longing for erotic encounters with a woman thirty years her senior represent three examples in question. All of these desires do not get approval from culturally-dominant paradigms of a good life; but there are people with such desires, and it is aggressive, violent, and paternalistic to approach them by claiming that they do not understand themselves in an appropriate manner because these types of acts do not correspond to the concept of good life, which the culture in question regards as valid. Initially, it was difficult for me, too, to imagine that it can be the case that a deaf person is not disabled but merely different. However, by recognizing the wide range of preferences, choices, tastes, and cultures in all parts of the world at various times, I came to realize how important it is to recognize that a different human being might regard different capacities as important and different shapes as attractive.

This approach has implications for dealing with the concept of family [2, p. 166-168]. A good example is the option of creating children with three biological parents. The UK is the first country worldwide which approved the necessary technologies for having such children. It is a very simple technology by means of which the nucleus of one female egg is removed and replaced by the nucleus of a different egg. In this way, the risk that a mitochondrial genetic disease is transferred from a mother to her child is excluded – the mitochondria are not in the nucleus but in the cytoplasm which surrounds it. Afterwards, the new egg is fertilized so that the child with three biological parents can come into existence. The UK has approved this procedure solely for mothers with this mitochondrial disease. However, it could also be a useful tool for a lesbian couple or two women and a man having a biologically related child. If three adults have a biologically related child this way, why should not they be allowed to marry, if this is their desire? Adults and a biological child are the basic constituent for a legal family in most cultures. However, this example reveals the enormous social implications, if posthuman perspectives are considered.

Self-overcoming and the good life (the relevance of a prolonged health span)

It is also possible to make claims valid for a great percentage of people. In this respect, a good health and a prolonged health span are widely shared qualities. Furthermore, it can be said that permanent self-overcoming is often identified with an increase of fulfilment. The following example helps in understanding this issue. Life as a child in a protected family is easier than life as a student. However, if you asked a student whether they would want to be a child again, most students would decline that offer, because they value the cognitive capacities, intelligence, and knowledge they possess. These goods are not solely positional goods, but they are also intrinsic goods which people do not wish to give up once they possess them. Similarly, as children cannot imagine what it would be like as students, students cannot imagine what it would be like as posthumans. Following this analogy, once we are posthuman, we would not want to be students anymore [2, p. 158-159].

Autonomous self-overcoming and heteronomous self-overcoming

It must be stressed that the technologies discussed should not be used to legally force people to use them in a certain way. We agree that it is a wonderful achievement to live in a liberal and pluralist society. Consequently, there are two options for using emergent technologies: by means of autonomous self-overcoming and by means of heteronomous self-overcoming whereby the second option does not allow political and religious leaders of other institutions to make a decision concerning who gets altered, but it refers to the very special parent-child relationship, in which such decisions are necessary and useful.

Gene technologies

Given the potential of CRISPR/Cas9, the so-called gene scissors which enables a precise, reliable and cheap way of modifying genes, I regard the future of human progress as closely related to a continued carbonate-based existence, which is particularly related to genetic research: gene creation by means of synthetic biology, gene modification, gene selection and analysis, and big gene data are four highly relevant fields in this context.

Synthetic biology is the attempt to create biologically-useful systems. Craig Venter's research is particularly notable in this field. He created a partially-synthetic species for which he is trying to get a patent which raises the ethical question concerning gene patents. He also claims to be the first to have created synthetic life. However, the Presidential Commission for the Study of Bioethical Issues holds that he has not created life. A closely-related field is that of biohacking which refers to a movement of do-it-yourself-at-home-biologists who conduct gene sequencing by means of easily-accessible methods and technologies. Both fields are thriving and it is worth getting further information on them.

A separate but related and much more intensely-discussed field is that of genetic modifications. For the past 15 years, it has been the subject of intense ethical discussions, and most leading ethicists in the world have taken a stance within this debate. The most noteworthy from a bioconservative perspective is the analysis and position Jürgen Habermas [15]. He regards genetic modifications for therapeutic goals as morally-legitimate, because, in this case, an all-purpose goal is promoted. The use of genetic modifications for enhancement purposes, on the other hand, is morally false, because, according to him, in this case persons get treated merely like objects. In addition, he presented reasons for rejecting a powerful pro-enhancement argument which analyses genetic enhancements and traditional education as structurally-analogous procedures. In both cases, parents make decisions for their offspring. Habermas claims, however, that genetic modifications are always irreversible and educational modifications are always reversible. Contemporary gene research and in particular epigenetics show that both premises are highly dubitable, if not clearly false. Consequently, there are strong reasons to claim that genetic modifications and traditional education are structurally-analogous processes. As structurally-analogous processes should also be treated as morally analogous, it may be concluded that both genetic and educational modifications can be morally-blameable- as well as praiseworthy – therefore, genetic enhancements do not have to be morally objectionable. This was a very short summary of an argument I have developed in great complexity and was published in 2015 in the Journal of Evolution and Technology [16]. It should also be noted that human genetic modifications cannot be yet reliably done on a day to day basis. However, scientific research and particularly the invention of CRISPR/Cas9 show that gene modifications can be undertaken precisely, reliably, and cheaply and that they can be successful without having side effects. Thus it is merely a matter of further research to turn this technology into a reliable one. I regard this issue as the most powerful technological invention of this decade. Furthermore, this issue will keep bioethicists particularly occupied in the forthcoming years.

Another separate issue is that of gene selection. It must be noted here that this is already a reliable technology. It presupposes in vitro fertilization which means that an egg gets fertilized by means of a sperm within a petri dish. After several divisions of the fertilized egg, it is possible to take one fertilized egg and analyse it genetically. Thereby, a great variety of information can be found, such as character traits, health information, and responsiveness to certain pharmacies. Then, it can be decided whether the fertilized egg from which the cell was taken should be implanted in the womb or not. Several issues are discussed in this context: What is the moral status of the fertilized egg? What is the status of the analysed cell which gets destroyed? Is it moral to select a fertilized egg after IVF and PGD (preimplantation genetic diagnosis)? Habermas responded in the same way as in the prior case: it is immoral, because a person is treated merely as an object, thus it is an immoral procedure. However, he does not recognize that fertilized eggs do not have to be identified with a person. For many people, a fertilized egg which consists of eight cells is merely a lump of cells and should not be identified with a fully developed human being. This position receives further support from the fact that 7 out of 10 fertilized eggs which come about by means of traditional procreation are never realized, but are eliminated with a slightly stronger monthly period. In addition, it needs to be understood that selecting a fertilized egg and selecting a partner for procreative purposes are structurally analogous procedures [17]. Just as a government should not interfere in one's selection of a procreative partner, a government should not interfere when we wish to select (or not) a fertilized egg after IVF and PGD.

Furthermore, gene analysis will be even more relevant than gene selection. Progress in this field is particularly noteworthy given the parallel creation of big gene data [18]. Such an analysis can tell you how likely it is for one to get a certain disease, it can reveal something about one's reaction to certain pharmacological products, and it can describe some of one's strengths and weaknesses. It is in our own interest to having such an analysis. On the basis of such an analysis one can change their life so that life goals are promoted by considering one's genetic dispositions. The costs of such an analysis are continuously getting lower. However, in this case, the issue of bioprivacy (or, more precisely, that of genetic privacy) arises – or rather the dissolvent of the concept of one's genetic privacy. The problem is already explicit. Let us consider the German legal system. It is already a legal obligation to give away the information of an already-made gene analysis, if you wish to make an insurance with a high financial risk for the insurance company. Consequently, anyone considering the option of having such an insurance, cannot realistically risk to have their genes analysed, although it could be in their personal interest to undergo such a gene analysis.

However, it is not only the person in cause who is interested in one's genetic information. Potential future employers, insurance companies, and the government could also be highly interested in this information. Once the data is available, it is digitally and hence publicly available, because we have reasons for holding that anything digitally available online is no longer private information, because it is already part of the internet panopticon. Still, the challenge goes even further, because one shares many central aspects of their genetic data with many of their family members. Even if one is not willing to give away this information, it is possible to get hold of the same data by asking their brother with whom they might not have spoken for years. Furthermore, one also needs to consider that by giving away genetic data, because of wanting to find information on one's family tree, one is giving away information about one's sister, too, although she did not permit this information to be out there. The question of genetic privacy represents significant further stepping stone concerning the future of technological developments.

In 2015, Kuwait has already made it legally obligatory for all citizens and residents to deliver tissue samples. If they do not or if they deliver false samples, they would face numerous years in prison. The official justification for this decision is fighting crime and terrorism. If successful in Kuwait, this legislation has the potential to be employed in more countries over the world. The technology is available, and many people are using it, because it provides us with important and fascinating information. *23andme*[1] has more than a million customers worldwide. Genetic information could be extremely useful for the future. However, gene analysis also has the potential of radically change the way societies are structured in the near future. Hence, genetic privacy is an issue which desperately needs to be discussed further by bioethicists, in particular the question on who should own the information of a gene analysis: Should it be me, because the information belongs to my genes? Should it

---

1 *23andme* is the first company to offer autosomal DNA testing for ancestry. See more: https://www.23andme.com/en-int/

belong to the analysis company, because they are mixing their labour with the genetic material, and mixing one's labour with something is the basis for acquiring property? Or maybe, the information should belong to all biological members of a family, as they are sharing a great amount of the information, and it could have negative implications for all of them?

The realms of gene technology and computer technologies merge at this point. We have big gene data which encloses important information about one's physiology and big IT data by means of which important psychological information is revealed. Furthermore, cartographies of all our movements are being digitally stored, by means of public surveillance and facial recognition software, telephone traces, and GPS data of in-car navigation systems. All three fields of data are digitally stored and processed on the internet, which again turns all the information into publicly available data. All central aspects of one's personality are revealed within the internet panopticon where citizens of all technologically-advanced countries are imprisoned. It is an unpleasant situation, which alters one's behaviour, once you are fully aware of your situation. However, a solution for this situation is far from clear.

Conclusion

This chapter provides a summary of the most important current and future bioethical challenges we are facing. Most of them belong to the field of gene ethics, as by means of these technologies we have the potential to radically alter the set-up of our societies and even of enhancing evolution – therefore, the likelihood of evolving into posthuman increases. This shift is not independent of a further ontological paradigm shift which has been occurring since the 19th century, and it is related to a revised self-conceptualization. We no longer regard ourselves as categorically dualistic entities which consist of a material body and an immaterial soul, but an increasing amount of human beings conceptualize themselves as non-dualist relational entities who are in a permanent process of self-overcoming. Consequently, we have to seriously consider that it would be naïve to assume that homo sapiens sapiens represent the final step of the evolutionary process. We need to realize that we will either die out or evolve further, as stagnation would necessarily lead to human extinction – if not now, then at least in 5 billion years when the sun will explode and destroy our entire solar system. This wider picture plays a central role in refocusing our debates in the field of bioethics.

References

[1] R. Ranisch, S. L. Sorgner. "Introducing Post- and Transhumanism." 7-29. In Ranisch, R./ Sorgner, S.L. *Post- and Transhumanism. An Introduction*, ed. R. Ranisch, S. L. Sorgner. Frankfurt a. M.: Peter Lang, 2014.

[2] S. L. Sorgner. *Transhumanismus: 'Die gefährlichste Idee der Welt'!?* Freiburg i. Br.: Herder, 2016.

[3] S. L. Sorgner. *Menschenwürde nach Nietzsche: Die Geschichte eines Begriffs.* Darmstadt: WBG, 2010.

[4] S. L. Sorgner. "Human Dignity 2.0. Beyond a Rigid Version of Anthropocentrism." *Trans-Humanities* 6(1) (2013): 135-159.

[5] P. Singer. *Animal Liberation.* New York: Harper, 2002.

[6] P. Singer. *Practical Ethics.* Cambridge: Cambridge University Press, 2011.

[7] D. B. Morton. "Self-consciousness and animal suffering." In *Biologist* (London)47(2) (2000):77-80.

[8] http://www.animalcognition.org/2015/04/15/list-of-animals-that-have-passed-the-mirror-test/, accessed: 11.02.2018.

[9] S.L. Sorgner. "Metahumanist Politics and Three Types of Freedom." In *European Journal of Science and Theology* 11(1) (2015): 119-129.

[10] Hayles N. Katherine. *Unthought: The Power of the Cognitive Nonconscious.* Chicago: University of Chicago Press, 2017.

[11] S. L. Sorgner. "Kant, Nietzsche and the Moral Prohibition of Treating a Person Solely as a Means." In *The Agonist* 4(1-2) (2013): 1-6.

[12] J. Savulescu. "Procreative Beneficence: Why We Should Select the Best Children." In *Bioethics* 15 (5-6) (2001): 413-26.

[13] J. Savulescu, G. Kahane. "The Moral Obligation to Create Children with the Best Chance of the Best Life." In *Bioethics* 23(5) (2009): 274-90.

[14] S. L. Sorgner. "The Renaissance Ideal, the Common-Sense-Ideal, and the Radical Plurality of Goodnesses The Transhumanist Types of (Post)Human Perfection." 141-157. In *Perfecting Human Futures: Transhuman Visions and Technological Imaginations,* ed. J. B. Hurlbut, H. Tirosh-Samuelson. Wiesbaden: Springer, 2016.

[15] J. Habermas. *Die Zukunft der menschlichen Natur. Auf dem Weg zu einer liberalen Eugenik?* Frankfurt am Main: Suhrkamp, 2001.

[16] S. L. Sorgner. "The Future of Education: Genetic Enhancement and Metahumanities." In *Journal of Evolution and Technology* 25(1) (2015): 31-48.

[17] S. L. Sorgner. "Is there a 'Moral Obligation to Create Children with the Best Chance of the Best Life'?" In *Humana Mente: Journal of Philosophical Studies* 26 (2014): 199-212.

[18] S. L. Sorgner. "Genetic Privacy, Big Gene Data, and the Internet Panopticon." In *Journal of Posthuman Studies: Philosophy, Media, Technology* 1 (2017): 87-103.

Can Ethics Govern Technology? Bioethics in the Age of Techno-Science

Corrado Viafora

Introduction

Bioethics has now become so well established that it is one of the most debated topics in our societies [1]. This is because - more as an expression of cultural sensitivity than as a specific discipline - it has succeeded in intercepting the ethical questions posed by the way in which our new powers of intervention on life are managed. These new powers are provided by science and technology, or rather by "techno-science" [2], a term that better describes the now virtually inextricable links that have been developing between science and technology. The ethical dimension of the questions raised by our handling of these new powers are self-evident, despite the techno-scientists' misplaced conviction that the issues can be solved in exclusively-technical terms. These powers carry not only benefits, but also associated risks, making it essential that we discuss together which and how many risks we are willing to take. These powers offer gains, but in many cases they also entail enormous costs, so we have to decide on the priorities that we intend to establish. Faced with the ambivalence of techno-science, with the awareness of its potential and also of the risks it carries when its powers touch on the innermost structures of life itself, what is directly called into play is our moral responsibility.

An excellent example of the type of new power we are talking about can be seen in the field of genetic engineering. It has the power to affect the structures that give all living organisms their specificity and individuality, and that combine together to produce the ecological balance of the whole biological system. After developing its ability to see a genetic structure like an organized dataset, genetic engineering has now enormously extended its capacity for knowing, changing, and using living organisms. It can be used to modify organisms to strengthen their resistance to certain parasites or herbicides, or to particularly adverse environmental conditions. We can generate transgenic animals as experimental models that biologically resemble humans, and possibly even use their organs for human transplants. We can produce synthetic bacterial genomes by assembling DNA sequences obtained in the laboratory and then inserting them in bacteria deprived of their own original genome in order to "train" them to have certain useful functions (in synthetic biology). We can alter the genetic sequence of the somatic and germinal cells of living organisms by means of gene editing technologies that are becoming ever more precise and focused, and also increasingly easy to use.

Even sketched in these few lines, this reference to the potential of genetic engineering suffices to show that the more technological interventions go into depth, and the more rapidly they advance, the more the ethical issues they raise become inescapable. For instance, is our current understanding of genetic and ecological dynamics sufficient for us to be able to predict the risks potentially deriving from our creating genetic combinations previously unknown to nature? The most convincing reason for taking this question extremely seriously has to do with the very concept of living organism. Unlike mechanisms, in which everything can be brought down to their single parts, and we can isolate single parts and contain our intervention on them, organisms cannot be reduced to the sum of their single parts, they are complex entities. Hence the difficulty of governing systemic effects, or in other words of controlling what happens to the whole when we take action on a part. We must not forget that genetic engineering enables us to redesign living organisms that are the product of three billion years of evolution. This fact in itself carries further uncertainties relating to how genetic engineering can change the speed of evolutionary processes. Caution is a first, mandatory condition for responsibly managing the new powers afforded by techno-science.

The increasingly insistent appeal for responsibility has become a general paradigm of ethics in the age of techno-science, but there is no doubting the fact that H. Jonas [3] takes the merit for placing the concept of *responsibility* at the very heart of our ethical reflections, even to the point of making it a guiding principle. The premise on which Jonas constructs his proposed "ethics for technological civilization" stems from the conviction that humans are capable of steering the new powers offered by technology towards certain goals. Basically, technology is seen as a tool – it may be as novel, complex and pervasive as you like, but it is still only a tool. Jonas explains this "humanist premise" in the following terms: "For the sake of human autonomy, the dignity of owning ourselves and not letting us be owned by our machine, we must check the technological stampede" [4] (pp. 891–898). Though the ontological foundations behind his approach may be open to criticism, its merits are undeniable. Jonas provides one the most lucid diagnoses that contemporary thinking has succeeded in expressing of the qualitative changes in human agency produced by technology. Above all, he is particularly convincing in claiming that the responsible management of technology is a fundamental moral imperative, an issue on which human dignity itself depends.

But in the light of the advances made in techno-scientific knowledge, is this "humanist premise" still tenable? In other words, is ethics capable of governing technology?

We cannot give this question any serious thought without discussing the lines of reasoning adopted by influential contemporary philosophers to justify their skepticism concerning the ability of ethics to steer technology. The theory that I intend to support is that a more critical approach to the relationship between technology and ethics, developing from some of the most representative traditions of our European culture, can discredit such skepticism and continue to sustain the credibility of the humanist premise.

Is ethics powerless in the age of technology?

Umberto Galimberti is the Italian philosopher who has contributed the most to the popularity of a line of thinking that - from Heidegger to Severino – claims to offer good reason for being sceptical, that he justifies with a particular interpretation of modernity [5]. He bases this interpretation – notoriously inspired by Heidegger – on the gradual identification of *reasoning* with *calculating*, typical of the evolution of rationality in the modern age. As *reasoning* has increasingly come to coincide with *calculating*, it is hardly surprising that technology has come to represent the strictest rationality. It follows that, as technology represents the strictest rationality, it is destined to absorb human rationality in all its facets, brushing away all its residual anthropological content (values, desires, passions). This becomes just as necessary as any natural process. Galimberti's peremptory conclusion is that, "Ethics as an indication of what 'must be' cannot avoid surrendering to technology, which *knows* how things must be in order to optimize well-being and growth according to criteria of the strictest rationality, based on an evolutionary pattern that increasingly resembles a *natural* process rather than a *historical* process" [6, p. 221].

Clearly, the sceptics' lack of faith in the capacity of ethics to govern technology knows no bounds. As Galimberti sees it, neither the *ethic of responsibility*, nor the *ethic of conviction* (the two extremes between which ethically-oriented human action oscillates according to M. Weber) stand a chance. The reason is simple: where technological action is self-perpetrating, it engenders consequences that are *independent* of any intention, and also *unpredictable* as regards their final outcomes. Faced with technology's ever-increasing independence and the structural unpredictability of its consequences, all we can do is realize the utter powerlessness of ethics. Referring particularly to bioethics, Galimberti says that:

> The ethic of responsibility demands that a subject be responsible not only for his intentions, but also for the effects of his actions. On one condition, however: that these effects be predictable. But

> it is characteristic of technology to have unpredictable effects - as in the case of cloning. This is where ethics has to chase after technology. So then what? It establishes that humans must not be cloned. But how can ethics ask technology not to do what it is capable of doing? Can it really do so? I think not [6] (p. 119).

But is it really true that technology wields such a totalizing power? If it were, and if we were therefore obliged to accept that ethics is completely powerless against technology, then all bioethics would have to do is serve as a sort of acclimatizing chamber, or incubator, to facilitate a process of gradual adaptation such that even outcomes that initially seem morally particularly troubling (like the case of cloning to which Galimberti himself refers) become acceptable in the end – where the term "in the end" simply means when ethics surrenders to the power of technology.

A more critical conceptualization of the relationship between ethics and technology

Going against this scepticism, the voices of some of the most representative traditions of contemporary philosophy are saying that we can see the relationship between ethics and technology in a different light. On the one hand, we can refute the theory according to which the process leading to the powerlessness of ethics is a natural necessity. On the other, we can create the right conditions for technology to be an element of moral creativity. These are two stances taken respectively by J. Habermas and J. Ladrière.

The conditions for bridging the gap between opportunity *and* capacity *to choose: Habermas's proposal*

Although modern societies do actually seem incapable of managing the knowledge and power that they increasingly have at their disposal, this powerlessness is by no means a natural necessity. It is the product of certain historical dynamics. This is what Habermas suggests, and this provides the starting point for a more critical conceptualization of the relationship between ethics and technology, that he approaches on the grounds of a different interpretation of modernity.

For Habermas, the dual project of modernity involves exerting a *scientifically* rational power and an *institutionally* rational power in order to grant a public debate regarding the way people are willing and able to live together. If this is the case, then we have to admit that this project has remained "incomplete". The reasons lie basically in the fact that, over the course of modern times, the rationality governing the powers gained by technology ("instrumental rationality") has become detached from the public debate on how to deploy these powers ("communicative rationality"). This has given rise to an ever-growing phenomenon that in some ways sums up our post-modern condition, and that finds its most evident and dramatic expression in the bioethical issues we face, i.e. the gap between our expanding *opportunity* to make a choice and our collectively weakening *capacity* to do so [7]?

So how can we bridge this gap? How can we strengthen our collective capacity to make a choice?

The evolution of modern Western society shows that certain dynamics gradually gaining ground are constantly eroding the forms and content of our traditions. In the past, it was the power of tradition that decided much of people's behaviour. But in modern society it is the individual who serves as the core reference value, and the areas in which he or she has freedom of choice have expanded considerably. The acknowledgement of an individual's personal autonomy as a value is certainly a step forward in the process of our emancipation, but it is not without its troublesome facets. The problems are particularly evident in the bioethical field, where managing our new powers of intervention on life can hardly be left to individual choice alone - if for no other reason than the fact that any usage of the new powers available in this field inevitably has consequences that extend beyond the individuals directly involved. They can affect everyone else's life too, and in some cases even risk manipulating the very structure of human socialization. So, where do we find the solution?

Habermas would answer that we need to develop our "communicative rationality," as the basis for a debate focusing exclusively on the power of the arguments. In societies like ours, which are the offspring of modernity, our ever-greater emancipation from the social norms dictated by tradition has undeniably made us freer. We come under fewer constraints. At the same time, we are now rather short of shared references. Argumentation thus seems to be the most valid way for us to decide together what rules should govern our living as a community. Between an objectively-founded decision and an arbitrary decision, there is space for making a choice in the light of an exchange of ideas in which reasons can be weighed, judged, discussed, criticized, improved, and accepted or refused.

During this debate, it is fundamentally important that everyone be willing to let the *strongest argument* prevail. That is why this practice can be qualified as "ethical." If we tend to let the argument of the *strongest speaker* prevail instead, then the line more or less explicitly adopted takes on a "strategic" nature. In practice, the former (ethical) solution demands that we create the right conditions for all potential participants in the debate to genuinely have the same opportunity to make their voice heard, and to exchange their views with those of others, constrained by nothing except the sole rational force of the best argument.

The theorists of deliberative democracy have dedicated particular attention to these right conditions, with which Habermas's ethic of argumentation is closely connected. Analyzing the social functions served by creating adequate spaces for debate in public health institutions, like ethical committees, Amy Gutmann (a well-known exponent of deliberative democracy) envisages that:

> A well-constituted bioethics forum provides an opportunity for advancing both individual and collective understanding. Through the give-and-take of argument, participants can learn from each other, come to recognize their individual and collective misapprehensions, and develop new views and policies that can more successfully withstand critical scrutiny [8] (pp. 40–41).

These indications are implemented in the deliberative practices of health care ethics committees, but they are equally applicable to the sphere of bioethics in general. There is an increasingly deep conviction that the proper management of bioethical issues demands a shift from individual decision-making processes to a more collective and participative approach.

Given this baseline condition, i.e. the commitment to getting more people involved, it might seem that making room for as many voices as possible to be heard will inevitably lead to an intensification of the moral conflicts. Experience has demonstrated that it is worth running this risk, however, given the benefits deriving from the fact that decisions that go through such a process enjoy a shared moral legitimization - the benefits of decisions made in the "first person plural" [8, p. 41].

The conditions for making the technological environment humanly habitable: Ladrière's proposal

Like Habermas, Ladrière recognizes the problems that ethics faces in our technological times. Habermas admits that ethics has come to seem powerless in the face of technology in our societies, but he also very convincingly demonstrates that this situation is not irrevocable. It is the outcome of clearly identifiable historical dynamics, and the precise aim of his critical analysis of these dynamics is to identify them clearly, and suggest how to combat them.

From his point of view (that of the philosophy of science), Ladrière relates the difficulty that ethics faces in the age of techno-science to a particular way of intending the influence of technology on ethics [9]. He sees technology as having a dual impact on ethics, that is *destabilizing* on the one hand, but *stimulating* on the other. Technology has a destabilizing effect because using the new powers of intervention on life that it offers uproots humans from the previously *stable* world they were used to dealing with, obliging them to cope with an unfinished reality that they themselves transform every day, and by which they too are transformed. The impact of technology is also stimulating inasmuch as the novelty and the potential it offers can serve as a stimulus for our moral imagination.

The category to which Ladrière refers in discussing our human being in the age of technology - and that he adopts as an ideal for regulating the relationship between ethics and technology - is "habitability". His use of this term makes it immediately evident that the core issue is the problem of how to manage not the potential of a given single technology, but our technological universe as a whole. So the fundamental question that Ladrière poses is how the artificial environment generated by technology can be humanly habitable.

His answer is inspired by an evocative phenomenology of habitability. For the technological environment to become humanly habitable, people need to have a sort of "familiarity" with it. It has to be a place where they can feel at home, where their "life can be conducted freely." According to Ladrière, the conditions needed for this to happen can be defined by the constitutive dimensions of the human being. For instance, a technological environment that generates a physically unsustainable sensory input would not be humanly habitable, nor would a technological environment that reduces human interactions to merely functional relationships. A technological environment that reduces the historical dimension of our being to the rhythm of a mechanism would likewise be humanly uninhabitable. Clearly, these are not conditions imposed by any abstract metaphysics, but derivable from the original experience that places human beings in contact with the life-world.

Based on this premise, the first condition for a technological environment to be humanly habitable has to do with our bodily dimension. It is with our bodies that we exist in direct contact with the world, inasmuch as concerns our *affectivity* (our capacity to feel, absorb, resonate), and our *effectivity* (our capacity to take action in the world that goes beyond the simple game of natural causality). The habitability of the technological environment should therefore be judged first of all from the point of view of its impact on our corporeality.

We must not forget, however, that our body mediates between our being and our world (which is why it is the first factor to consider in determining the habitability of the latter), but only if it is part of a shared space for interaction. A space where people constantly meet, create meanings for themselves, and exchange them with others - to such a degree that the human being can only fulfil itself through *coexistence.* As a consequence, another equally important parameter for assessing the habitability of our technological environment consists in its impact on the community of human beings - on their coexistence – rather than on any given particular situation.

But then our being, considered in its corporeal dimension and in the sense of our coexisting with others, is experienced in a structural *tension* marked by a constant sense of incompleteness – an ontological condition takes the form of a *lived temporality*. This is a totally different matter from the physical temporality in which the scientific representation of the world places all of its phenomena. Lived temporality is the condition that turns our existence into the story of its manifestation, making it both heir to itself and the promise of its future. It is another essential parameter against which to assess the human habitability of the technological environment.

In short, this environment - in which increasingly large proportions of our lives are enacted - can be considered humanly habitable to the degree in which it guarantees the conditions that enable us to keep a grasp on our humanity: our bodily integrity; our participation in each other's lives; and our opportunity to have a story that has a sense.

All this is certainly hugely important, but does not completely cover the heuristic breadth of the category of habitability, which has to do not only with "compatibility", but also with "exchange and reciprocity." Ladrière attributes a special importance to this aspect, given the significance that he assigns not only to the destabilizing impact of technology on ethics, but also to its value as a provocation and a stimulus.

As a consequence, one condition that must certainly be met for the technological environment to be in harmony with our human being is that it provides a field in which our constitutive dimensions can develop. But another equally important condition is that our being can be enriched by the new possibilities offered by technology. If our technological environment succeeds in integrating our being, then a sort of resonance is produced that enables our being to chart its course in this environment, and enables the technological environment to add novel meanings to our being, opening it up to hitherto unexplored pathways.

Of course, when it comes to dealing with practical situations, the approach dictated by this fundamental habitability criterion can only serve as a general guideline. But, as Ladrière reminds us, in the uncertain and sometimes ambiguous and ambivalent circumstances in which human actions are undertaken in the age of technology, valid support can come more from an ethical awareness of the complexities of the situation than from norms that strive to cover every particular issue.

Conclusion: for a bioethical discourse capable of integrating the normative and the hermeneutic dimensions

This albeit brief presentation of the proposals advanced by Ladrière and Habermas should suffice to illustrate the conditions needed for us to effectively overcome any scepticism concerning the ability of ethics to steer technology.

The strength of Habermas's proposal lies in that it reveals the ideological element inevitably contained in the assumption that ethics is *naturally* powerless in the face of technology. It is ideological in the sense that this assumption makes us tend to overlook the real historical dynamics behind the apparent powerlessness of ethics, and to neutralize any well-argued, transparent and participative public debate on the subject.

The strength of Ladrière's proposal consists in its original interpretation of the relationship between ethics and technology. It justifies not only the capacity of ethics to steer technology according to the need for the technological environment to be humanly habitable, but also - and equally significantly - the role of novel technologies in stimulating our moral imagination.

In addition to demonstrating the general feasibility of ethics governing technology, Habermas and Ladrière also suggest how to proceed in order to deal adequately with the problematic sphere of bioethics as a whole.

Here the problem has to do with two fundamental aspects. The first concerns the *normative* domain, i.e. which principles and rules should govern how we solve the novel conflicts raised by the need to manage our new powers of intervention on life responsibly. The second concerns the *hermeneutic* domain, and how we interpret the human sense of our being in the face of the provocations raised by novel technologies.

As mentioned earlier, the approaches taken by Habermas and Ladrière point in two directions: the former towards creating the right conditions, in an increasingly pluralist cultural setting, for the development of norms that are as universally applicable as possible; the latter towards inscribing the technological environment within the effective motion of our being, and this – given the original experience that precedes every representation - has to do with corporeality, with otherness, and with temporality.

Leaving aside the specific means proposed by the two philosophers, what they are both saying is that, if we want to address the whole array of bioethical issues, then both the above-mentioned dimensions - the normative and the hermeneutic - have to be involved in substantiating the bioethical discourse.

References

[1] A. Jonsen, *The Birth of Bioethics*. New York: Oxford University Press, 1998.

[2] G. Hottois, "La technoscience: de l'origine du mot à son usage actuel", *Regards sur les technosciences*, J.-Y. Goffi, ed., Paris: Vrin, pp. 21-38, 2006.

[3] H. Jonas, *Das Prinzip Verantwortung. Versuch einer Ethik für die technologische Civilisation*. Frankfurt a.M.: Insel, 1979. Engl. tr. *The Imperative of Responsibility: In Search of an Ethics for the Technological Age*. Chicago: University of Chicago Press, 1984.

[4] H. Jonas, "Technology as a Subject for Ethics", *Social Research*, vol. 49, no. 4, pp. 891-898, 1982.

[5] U. Galimberti, *Psiche e techne*. Milano: Feltrinelli, 1999.
[6] U. Galimberti, "La tecnica e l'impotenza dell'etica", *Etica del plurale. Giustizia, riconoscimento, responsabilità*, E. Bonan and C. Vigna, eds., Milano: Vita e Pensiero, pp. 201-221, 2004.
[7] J. Habermas, *Moralbewusstsein und kommunikatives Handeln*. Frankfurt a.M.: Suhrkamp Verlag, 1983. Engl. tr. *Moral Consciousness and Communicative Action*, Cambridge, MA: MIT Press, 1983.
[8] A. Gutmann and D. Thompson, "Deliberating about bioethics", *Hastings Center Report*, vol. 28, no. 3, pp. 38-41, May-June 1997.
[9] J. Ladrière, *L'éthique dans l'univers de la rationalité*. Namur: Artel; Québec: Fides, 1997.

The Moral Relationship of the Human and the Non-Human Animals in Light of Ethology

Alexander Krémer

Introduction

The relationship between animals and human beings is a complex problem. Especially if one perceives it within a historical perspective. This chapter will not undertake the monographic analysis of the problem, but will approach the theme through animal experiments. Besides animal farming (or factory farming), animal experiments show perhaps the most, what sort of pain humans can cause animals.

It is beyond question that millions of people purchase cosmetics, medicine, and household items day by day. It means that everybody uses products whose effects have previously been tested on animals. However, not only the significant cosmetic companies, chemical and pharmaceutic factories conduct so-called animal experiments, but also military laboratories (first of all those, where chemical weapons are created and tested), and the research institutes of the medical and veterinary universities. Whether it is called "animal testing" or "animal research," it refers to the experimentation carried out on animals. These are "animal experiments."

The animal rights movement – Singer and Regan

The animal rights movement (also known as the animal liberation movement, animal personhood, or animal advocacy movement), is a social movement which tries to end the rigid moral and legal distinction drawn between human and non-human animals. It also tries to end the status of animals as property and to end their use in research, food, clothing, and entertainment industries. It is one of the rare cases when a social movement was created which is, to a large extent, sustained academically by philosophers. The movement was allegedly founded in the UK in the early 1970s by a group of post-graduate philosophy students, by the so-called "Oxford Group," but it was strengthened by Peter Singer (utilitarian liberationist) and Tom Regan (animal rights advocate).

Peter Albert David Singer is an Australian philosopher who taught applied ethics at different universities. He specialized in applied ethics and approaches ethical issues from a utilitarian perspective. Singer is best-known for his book *Animal Liberation*, regarded as the cornerstone of the animal liberation movement [1]. Not all members of the animal liberation movement share this view, and Singer himself has said the media overestimates his status. His views on animals and other issues in bioethics attracted him attention, as well as a degree of controversy.

Animal Liberation, published in 1975, has been cited as a significant influence on leaders of the modern animal liberation movement. The central argument of the book is an expansion of the utilitarian principle that "the greatest good for the greatest number" is the only measure of good or ethical behaviour. Singer argues that there is no reason not to apply this to other animals. He made the term "speciesism" popular, which was originally coined by Richard D. Ryder to describe the practice of privileging humans over other animals. In *Animal Liberation*, Singer argues against speciesism on the grounds that a being belongs to a certain species. He holds the interests of all beings capable of suffering to be worthy of equal consideration, and that giving lesser consideration to beings based on their species is no more justified than discrimination based on skin colour. He argues that animals should have rights based on their ability to feel pain more than their intelligence. He says, in particular, that while animals show lower intelligence than the average human, many severely intellectually challenged humans show equally diminished, if not lower, mental capacity. What is more, some animals have displayed signs of intelligence (for example, primates learning elements of American sign language and other symbolic languages) sometimes on par with that of human children, and therefore intelligence does not provide a basis for providing nonhuman animals any less consideration than mentally disabled humans.

Singer does not specifically contend that humans do not have to use animals for food insofar as they are raised and killed in a way that actively avoids the inflicting of unnecessary pain, but as such farms are uncommon, he concludes that the most practical solution is to adopt a vegetarian or vegan diet. Singer also condemns vivisection except where the benefit (regarding improved medical treatment and others) outweighs the harm done to the animals used.

Tom Regan is an American philosopher who was specialized in animal rights theory. He taught from 1967 until his retirement in 2001 at North Carolina State University. Regan is the author of numerous books on the philosophy of animal rights, including *The Case for Animal Rights* (1983), a book in a collection of studies which have significantly influenced the modern animal rights movement [2]. He argues in these studies that non-human animals are what he calls the "subjects-of-a-life," just as humans are. Furthermore, he argues that if we want to ascribe value to all human beings regardless of their ability to be rational agents, then we should be consistent and similarly ascribe it to non-humans. Regan argues in *The Case for Animal Rights* that non-human animals bear moral rights. His philosophy aligns broadly within the tradition of Immanuel Kant, though he rejects Kant's idea that respect is only due to rational beings. Regan points out that we routinely ascribe inherent value, and thus the right to be treated with respect, to humans who are not rational, including infants and the severely mentally impaired. The crucial attribute that all humans have in common, he says, is not rationality, but the fact that each of us has a life that matters to us; in other words, what happens to us matters to us, regardless of whether it matters to anyone else. In Regan's terminology, we each experience being the "subject-of-a-life." If this is the true basis for ascribing inherent value to individuals, he argues, in order to be consistent we must ascribe inherent value (hence, moral rights) to every subjects-of-a-life, whether human or non-human.

The common precondition of Singer and Regan's theory

Singer and Regan's argumentation has a common feature. They both refute *rationality* as the basis of morals and morality. They both believe that rationality is unacceptable as the basic criterion of becoming a moral agent. However, in my opinion, without rationality, we cannot speak about morals and morality. Rationality is a condition sine qua non, an indispensable condition of the moral phenomenon. It follows from this that *animals cannot become moral agents!*

Why can we state that animals cannot be moral agents? In order to answer this question, we first of all need argumentation not taken only from the history of philosophy, but from science as well. The philosophical argumentation *in itself* cannot satisfy our needs of verification because it can clarify mostly the standpoints within intellectual history, their relationships, and the contemporary scientific, theological and other views, which determined them. It is beyond question that the issue mentioned above needs a philosophical and ethical approach too. However, the exclusive exegesis of the earlier philosophical and ethical texts can mostly lead to obsolete results if we do not consider recent scientific recognitions. Thereby, I am arguing that not only an approach from the history of philosophy is needed, but also philosophical and ethical trains of thought, which are proved by recent results and argumentations of the positive sciences.

Rationality as a necessary precondition of morals and morality – two arguments (choice and ethology)

There are at least two clear arguments for my standpoint. The first one is a logical and experiential argument, and it goes mostly against Regan. The second one is a historical and ethological argument, and it goes mostly against Singer. According to the first argument, a moral agent must be rational since a moral agent has to know what is good and bad in a moral sense. Without this sort of knowledge, we cannot speak about a moral decision. *The moral choice is always rational.* Most people grow in a human community where some moral norms already exist, and the youngsters appropriate the basic moral norms and values together with their native tongue and community traditions.

According to my second argument, *morals and morality are exclusively socio-historical products.* They were born in a particular phase of mankind's historical development when some sort of rationality had already existed. This "historical moment" is, of course, a rather long period of a historical transition, where moral norms were born from customs and taboos. Ethology comes here into picture because it is attached to my second argument.

Why can animals not become moral agents? As I have already mentioned, we first of all need scientific research and argumentation and not philosophical argumentation since the latter speaks only about the philosophers' standpoints of their times. What is more, an *interdisciplinary research* would be needed with the help of zoology, biology, sociology, sociobiology, neurology, evolutionary psychology, ethology, human-ethology, and so on.

Due to our limits, I mention here only ethology and human ethology. *Ethology* is a branch of knowledge dealing with animal behaviour and human character under natural conditions, considering their formation and evolution. Ethology is an evolutionary science, and *its basic principle* is that if we discover the evolutionary history of species, then we can also understand and explain the basic forms of their behaviour. This principle led to the creation of *human ethology* (based on Charles Darwin's, Konrad Lorenz's and Niko Tinbergen's oeuvre), which science *examines the biological basis of human behaviour, with particular attention to its evolutionary preconditions.* From the three main trends of the human ethological research [a) human capabilities in primates; b) research on babies and toddlers; c) comparison of different human cultures], the most important for us is the first one: the comparison between humans and primates.

As professor Vilmos Csányi, a well-known Hungarian human ethologist says, "*the human being is only a primate from a biological point of view*" [3]. Furthermore, Csányi also says that when comparing humans and animals it is forbidden to detach elements from the whole context of the evolutionary process. It is *a human behavioural complex*, which has – according to Csányi – three constitutive parts:

1) new behavioural forms connected to the life of a group;
2) harmonization of the group activities;
3) actions and capabilities of construction. [3, p. 301]

The new behavioural forms are connected to the life of the group, and the most important evolutionary result is that a permanent and closed group structure was born. This evolutionary happening has several preconditions: settlement; decreased aggression in sharing food and sexuality; new forms of division of labour; mental representation of the group, and loyalty to the group. However, the most noteworthy fact is that these features cannot be found in primates.

The harmonization of the group activities mostly refers to the ability to organize a group and create new forms of synchronization. Its preconditions are empathy (synchronization of emotions), imitation (synchronization of behavioural patterns, which creates the possibility of teaching), disciplining, and the ability to follow rules. After a while, the synchronization of activities, emotions and behavioural patterns happens *at the same time*, not only through language but also through creating music, singing, dancing, rites, pictures, etc. These features cannot be found in primates.

The actions and capabilities of construction happened and developed in the following fields:

a) the usage of language (animal language is a closed system, but human language is an open system with almost infinite possibilities of creativity, which allows the mental representation of anything, conversation about anything, and abstract thinking)
b) we also discuss the construction regarding the usage and creation of tools (it promotes and strengthens a. and c.)
c) abstract thinking (the "creation" of time: past-present-future; the creation of ideal worlds, which allows the existence of mathematics, religion, arts, etc., and also morals).

It must be emphasized that the developments of these abilities of construction *promote and strengthen each other!* Furthermore, *the group-structure, customs, taboos, technical and legal norms, etc.* are also results of construction! It is beyond question that these capabilities of construction exist in animals only in proto-forms! [3, pp. 301-309].

Summarizing this ethological description, it should be highlighted that these remarks show only a small aspect of ethology and human ethology. However, it must be emphasized that these features of the human behavioural complex are intertwined and in mutual relationships. In Csányi's opinion, the most essential aspect of this is that the strong group-structure, the capacity of synchronization, and the activities of construction result, together, in a closed feedback loop. The reason for that is that *the activities of construction are directed mostly at the group itself*, and the synchronization activities strengthen its effects, and loyalty to the group makes it permanent. The most important consequences of these facts are that a) *humans are culture-creating beings*, and b) *humans are determined by both biological and cultural evolution.*

Returning to our question ("Why cannot animals become moral agents?"), the following five conclusions can be drawn:

- Morals and morality are *not identical* with their biological basis, as the roof and the walls are not identical with the foundation of the house.
- *Morals and morality are exclusively socio-historical products.* Neither moral norms, nor the concept of "moral good" existed in pre-historic times! It is provable that the Greeks were the first ones to discuss morals on the social level, and Aristotle wrote the first work dedicated to ethics, the *Nichomachean Ethics.*
- Human beings became moral agents through *rationality*, not through the ability to suffering. Furthermore, humans do not maintain the moral phenomenon through the ability to suffer, but through *rationality.*
- That is why Singer's idea of replacing rationality with ability to suffer cannot be a accepted
- *Morals and morality are the results of human rationality*, and are exclusively socio-historical products. Therefore Kant's idea is able to describe the relationship between humans and animals.

Kant's moral theory on the relation between humans and animals

Before summarizing Immanuel Kant's view on the relationship between humans and animals, let us show the main structure of his ethical system. The most important issue in Kant's ethics refers to what is missing, that is, God. I am persuaded that it is a secularized version of the Christian moral philosophy, and in Kant's case, morality is identical to rationality. According to traditional Christian ethics, every moral norm and principle is deducted from its ultimate basis, God. It is a deontological model of ethics since people believe that there is an Absolute, which is also the highest moral good -- thus, only this can lay the absolute foundation for morality. Although Kant says in the preface of the *Critique of Pure Reason* (2nd edition) that "Thus I had to deny *knowledge* in order to make room for *faith*," [4, p. 117], it only signifies the result of his method of criticism. In his ethics, God is replaced by Reason, and God functions as a mere postulate. The categorical imperative is namely a synthetic, a priori proposition since it is created exclusively by the pure reason for the practical reason: "Act only according to that maxim whereby you can, at the same time, will that it should become a universal law" [5, p. 71]. Furthermore, we can find the same importance of rationality in moral situations. What should we do to become moral in moral situations? We have to test our subjective, individual maxims (only this way, they can become absolute moral laws) on the categorical imperative, which does not contain any particular moral value. (The fourth formulation of the categorical imperative is an exception in some sense [5, p. 87]). The categorical imperative determines only a relation between our maxims and a universal moral law, and we have to understand this relationship and act for the sake of duty, which means the respect of the categorical imperative. If our action fulfils the categorical imperative only accidentally, then our action will be *legal* but not *moral*. It means that we have to act in an absolute rational way if we want to become moral persons. (From a moral point of view, we can find in Kant's system the subjective maxims, the objective and absolute moral laws, and the categorical imperative.) We may not forget, however, Kant's imperative, which states that we cannot consider any of our emotions, desires or inclinations in a moral decision[1] [6].

Returning to our topic, it must be emphasized that Kant, following Aquinas, opposed the idea that humans have direct moral duties towards nonhumans. For Kant, cruelty to animals was wrong only because it was bad for humankind. In 1797, he argued that

> violent and cruel treatment of animals is far more intimately opposed to man's duty to *himself*, and he has a duty to refrain from this; for it dulls his shared feeling of their pain and so weakens and gradually uproots a natural predisposition that is very serviceable to morality in one's relations with *other men* [7, p. 238].

Kant's philosophy is loaded with several self-contradictions, but it seems that he is correct in connection to our moral obligations towards animals. Namely, in his opinion, humans do not have *direct* moral obligations towards animals since animals are not rational beings. Such beings cannot exist as moral agents because they cannot make moral decisions, which are impossible without the knowledge of the moral good and bad, and the awareness of their difference. This does not mean that in Kant's opinion, one can commit anything towards animals, as he thought that we always have *indirect* moral obligations towards them.

It means that contrary to most human beings,[2] animals are not moral

---

[1] This part of Kant's ethics is highly problematic, but I focus here only on one main issue: what is Kant's opinion about the moral relationship between the human and non-human animals, and why it is better, than Singer's and Regan's standpoint. The analysis of these other problems would need a separate article or monograph. However, I wrote more about this issue in my book: Krémer Sándor, *Alapvető etika* (Szeged: JatePress, 2004).

[2] For example, human babies are not moral agents since they are not rational enough or at all. Incompetency is a similar case, which shows that not only in the field of morals,

agents, but we, humans also have to be moral towards animals in order to maintain our human dignity. Every human being who is able to comply with this moral request, has already satisfied not only his or her duty to behave morally towards animals, but most importantly, the duty to maintain his or her human dignity.

References

[1] Regan, Tom, *The Case for Animal Rights.* University of California Press, 1983.
[2] Singer, Peter, *Animal Liberation.* New York: Harper Collins, 2002.
[3] Csányi, Vilmos. *Az emberi természet biológiai gyökerei.* Lecture at the Mindentudás University. Available online here: http://web.t-online.hu/csanyi14/Csanyi_Vilmos/Letöltés_files/mindentudas_egyeteme.pdf, 301. - CS-MIE (last time accessed: April 13, 2018).
[4] Kant, Immanuel, *Critique of Pure Reason.* (The Cambridge Edition of the Works of Immanuel Kant) Cambridge: Cambridge University Press, 1998.
[5] Kant, Immanuel, *Groundwork of the Metaphysics of Morals.* (A German-English Edition) Cambridge: Cambridge University Press, 2011.
[6] Krémer Sándor, *Alapvető etika.* Szeged: JatePress, 2004.
[7] Kant, Immanuel, *The Metaphysics of Morals.* (Texts in German Philosophy) Cambridge: Cambridge University Press, 1991.

---

but also in the criminal law, those human beings cannot be moral and legal agents, who are not rational enough, and I could enumerate other examples as well.

PART II

ETHICS OF ENHANCEMENT TECHNOLOGIES

Neuroethics and Moral Enhancement: The Path to a Moral World?

Maria Sinaci

Introduction

As imperfect beings, people have always had the desire to improve their condition: be healthier, be more visually appealing, live longer, be more intelligent and even be kinder, that is, more moral. Because of the scientific progress of the past few decates, these aspirations can become, at least partially, a reality. The purpose of new biotechnologies is not solely curative and aimed at the treatment of major medical conditions or disabilities; new biotechnologies now also aim human enhancement. Science now offers possibilities which were unthinkable a few decades ago with respect to the enhancement of certain characteristics of the human nature – possibilities which allow people to be "better than well" [1].

The concept of human enhancement is increasingly present in the academic environment, although it is used in a variety of methods – which leads to the conclusion that it is not clearly understood. Some specialists use it when referring to interventions which increase or intensify certain capacities. Others are more restrictive and use it exclusively for the description of those interventions which go beyond the medical field – thus, they create a border line between therapy and the extension of human capacities beyond therapy. In *Beyond Therapy* [2], the President of the Council on Bioethics defined enhancement in contrast to therapy, arguing that enhancement is the usage of biotechnological power to modify, through direct intervention, not the processes of disease and illness, but the "normal" functioning of the human body and mind, in order to increase or improve their native capacities. Chatterjee [3] makes the same distinction, arguing that biomedical enhancement refers to interventions used for the enhancement of certain capacities beyond normal, rather than re-establishing certain functions or capacities lost because of illness. In a functional-argumentative approach, "interventions are considered enhancements … insofar as they [augment] some capacity or function (such as cognition, vision, hearing, alertness) by increasing the ability of the function to do what it normally does" [4, p. 2].

Biomedical human enhancement aims different zones, such as cognition, memory, certain physical abilities or personality and behavioural traits, thus generating intense debates on the ethics of enhancement. However, the most recent challenge of ethicists is represented by moral enhancement, proposed by some as a solution to the serious moral problems of humanity. J. Săvulescu şi I. Persson [5] [6] argue for the necessity to enhance human moral behavior, because, they say, people are not psychologically and morally prepared to face today's global problems. Therefore, they affirm the necessity of moral enhancement methods such as using pharmaceutical products, neurostimulation, genetic editing, or any other type of biological method which can be used for this purpose. Generally, there is a hope within those who support moral enhancement that new biotechnologies could contribute to the construction of a better world through the enhancement of moral reasoning, the increase of pro-social behavior, and the strengthening of moral virtues. It would seem that the benefits of a society composed of as many moral people as possible are evident. But would this be possible? Is this the path towards a moral world? Is morality an objective for all people?

Before answering these questions, certain conceptual observations are necessary, which is why the following section will discuss the terminology on morality, moral enhancement, and neuroethics.

A terminological frame

In a wide sense, morality may be defined as "beliefs about what is right behaviour and what is wrong behaviour" [7]. However, this definition is somewhat relative and ambiguous. Establishing what is right and wrong in the behaviour and actions of a person is a central theme within debates on the standards of morality. Bernard Gert [8] is sceptical on having a sole definition of morality which could be applicable to all moral discussions – this is because morality seems to be used in two wide senses: descriptive and normative. Morality is linked to behaviour and authors such as T. M. Scanlon [9] and S. Darwal [10] highlighted the social nature of morality. The intentions and behaviour of each moral agent affects the well-being of others. Therefore, a larger number of moral invididuals in a community would increase the well-being of that entire community. The growth of the level of morality is not a new preoccupation – it has been present, in a specific form, in the religious sphere since Antiquity, and in monotheistic religions. The novelty in moral enhancement stands in the way it can be realized, through the most recent scientific descoveries such as neurobiology, neuropsychology, neurosociology, neurosciences, pharmacology, and of course medical biotechnologies. The ethical and philosophical implications which appeared within this context led to the birth of a new field, namely neuroethics.

The term "neuroethics" was used initially to indicate certain ethical aspects present in clinical neurology and brain research. Later on, the field and the procedures of neuroethics extended, enlarging its sphere of competence. The introduction and the establishment of the term is attributed to William Safire who defined it as "the examination of what is right and wrong, good and bad about the treatment of, perfection of, or unwelcome invasion of and worrisome manipulation of the human brain... [neuroethics] deals with our consciousness – our sense of self – and as such is central to our being" [11, p. 5]. In her article *Neuroethics for the new millennium* [12], Adina Roskies, neurologist and philosopher, proposes an approach of neuroethics in two divisions: the ethics of neuroscience and the neuroscience of ethics. The first division, the ethics of neuroscience is neuroethics as applied ethics. It investigates the ethical implications and questions arised in the research and usage of neuroscientific technology; furthermore, it also investigates whether its applications are in conformity with ethical values and principles. The neuroscience of ethics investigates empirically the biological basis of ethical thought and behaviour and the way in which it could influence ethical thought, as well as the way in which the brain generates thought and moral action. One of the topics of neuroethics, for instance, refers to the way in which intuitions are generated. Which parts of the brain are involved and how does the brain work in order to make a moral agent believe that a particular action is forbidden, allowed, or mandatory? Furthermore, it interprets philosophical notions such as free will, personal identity, intention, altruism, and value judgement.

The study of the brain and the new neurotechnologies have stimulated researchers towards the study of morality, especially concerning the human behaviour and emotional dispositions. But how could one define moral enhancement? What would being "more moral" mean? Several definitions exist on moral enhancement, but there is no consensus on them. Some definitions concentrate on moral behaviour or the effects of behaviour, while others concentrate on the process of moral decision-making, on human virtues, or on cognitive and affective capacities. According to T. Douglas [13], there are interventions which, in an anticipatory way, will give an individual more moral reasons or behaviours than (s)he would otherwise have had. Karim Jebari [14] understands moral enhancement as an improvement of empathy. I. Persson and J. Savulescu [15] consider that moral enhancement encompasses several factors, such as emotions, empathy, and reasoning. A more precise definition from the perspective of how one can intervene for the purpose of moral enhancement is given by B. Earp și J. Savulescu [16]. The authors discuss moral neuroenhancement which acts directly on the brain and propose the following definition: "Any change in a moral agent—effected or facilitated in some significant way by the application of a neurotechnology—that results, or is reasonably expected to result, in the agent being a morally better (i.e., more moral) agent" [16, p. 4]. H. Wiseman proposes a more concise definition for biomedical moral enhancement as following: "Some technological or pharmacological means of affecting the biological aspects of moral functioning, to boost what is desirable, or remove what is problematic" [17, p. 6]. However, beyond all the nuances of these definitions, the collective element in them is the scope of moral enhancement: the correct moral behaviour.

Is there a biological basis of morality?

The studies on the biological basis of thought and ethical behaviour have inevitably led to the question of whether there exists a biological basis of morality. In fact, the question regarding the origin of morality has existed for centuries in the thoughts of philosophers, theologians, and not only. The development of neurosciences, biology, psychology, and genetics, has allowed the empirical approach of certain questions, thus leading to the development of a naturalist frame capable of explaining the biological origin of moral behaviour. The research of the past few years, which explores the connections between the brain and morality, prove that the idea of a biological basis of morality has to be taken into consideration.

The most prominent supporter of the idea that morality is rooted in the biology of the brain is Patricia Churchland [18, 19]. Churchland [18] argues that morality is formed in a neurobiological frame which was submitted to evolutionary and cultural pressures, and modelled through interconnected cerebral processes, which eventually led to moral behaviour. The dimensions of this neurobiological frame are the following:

a. Care – rooted in the biological attachment towards family and friends and the care for their well-being
b. The ability to recognize psychological states – rooted in the benefits coming from the anticipation of the behaviour of others
c. Resolving issues in a social context
d. Learning social practices (positive and negative)

It is unclear however if the dimensions of P. Churchland are sufficient to answer all aspects of ethical thought. The essential element of moral behaviour is considered to be social interaction. Beyond this however, it seems that our moral behaviour is influenced by biological factors. Therefore, the studies which explore the connection between the brain and morality have led to interesting results: the levels of oxytocin and vasopressin influence high-level social phenomena such as trust, generosity, or punishment. Although Churchland demonstrates that a certain hormone, oxytocin, stimulates pro-social moral behaviour, it does not mean that we can conceive morality as being innate and we cannot, at least for the moment, connect it to a specific gene. However, the argument of the author that morality has biological roots proves that our moral behaviour is not solely motivated by reason. A research conducted by J. Green et al. [20], done by using functional MRI, has demonstrated the existence of cerebral areas which become active when the subject is involved in the evaluation of moral situations or even when his/her answers are lies.

Other researchers have also focused on this subject and argue for a biological basis of morality. Thus, the studies of an American group of researchers, demonstrated that moral judgement is influenced by genes which produce or regulate the neurotransmitters of each neural system involved [21]. The results suggest that the variation in the oxytocin receptor gene (OXTR) is associated with differences in moral judgement. According to the authors, "these data provide the strongest evidence to date linking variation in a specific gene to variation in moral judgment.."[21, p. 6]. The research done with the help of functional MRI have shown that "true" or "false" arguments activate different areas of the prefrontal cortex [22].

Research has shown that there is a link between moral thought and the biological aspect, but a gene responsible for morality is yet to be identified. Despite this, studies demonstrate that, on a cerebral level, neural systems allow for a moral judgement, based on neurotransmitters produced by certain genes. Interventions on neurotransmitters have been proven to have significant influence on moral judgement. All this leads to the fact that a biological basis of morality needs to be taken into consideration. If a moral biological basis exists, then moral enhancement is possible by stimulating/inhibiting the biological elements involved. However, it remains to be discussed what exactly influences moral enhancement, moral knowledge, motivation, or moral behaviour. Nevertheless, we consider that the specifics and particularities or morality are not given by biology. What biology offers, especially through the action and adjustement of neurotransmitters, are certain tendencies towards action, such as a sense of correctness, empathy, care for others, respecting rules – in other words, the general ability to develop a moral system. The specific rules we apply in our society are not given by biological data, but are given by moral agents.

Types of moral enhancement

The promoters of moral enhancement (most of whom consider traditional moral education to be inefficient) argue that biotechnological means should be used in order to increase the level of morality and strengthen humans from a moral point of view. Their argument is mostly related to the benefits this would bring to communities. If the genetic and biological influences on decision-making and morality are well understood, their enhancement with with certain products and technologies could be considered. Accoding to recent studies (V. Dubljevic şi E. Racine [23]), direct interventions on moral enhancement are divided in two types: neuropharmacological interventions and neurostimulation interventions. Studies have shown that moral judgement is produced through the coordinated interaction of several neural systems, each of which is based on a characteristic set of neurotransmitters whose adjustement can influence moral judgement. Interventions on neurotransmitters can have significant influence on moral judgement. For instance, oxytocin (also known as the "moral molecule") raises pro-social behaviour such as empathy, altruism, cooperation, and trust within the social group [24, p. 150, 190]. Research also suggests that variation in the OXTR (variation in the oxytocin receptor gene) most likely influences moral judgment by modulating the influence of oxytocin in the neuron, pathways that have previously been identified as preferentially supportive of ethical moral judgment [25].

Serotonin, another neurotransmitter, modifies moral judgement and behaviour by increasing the subjects' aversion towards harming others. Therefore, it could promote pro-social behaviour. Once morality is demonstrated to have, at least partially, a biological support, it could become influenced by pharmacological interventions. Taking as example the aggressive behaviour, studies in genetic behaviour and neurosciences have shown that a biological basis for violence does exist [26]. There are drugs which increase the control of impulse and can thus contribute to the moral behaviour of people. Ritalin is the drug which reduces violent impulses and contributes to moral enhancement by increasing the capacity to concentrate and think more deliberately on how to resolve issues and conduct ethical analyses. Modafinil also increases pro-social behaviour, such as empathy, cooperation, and trust [27, p. 2, 3]. Antidepressants (SSRIs - selective serotonin reuptake inhibitors) generally reduce aggression and seem to make people more cooperative, sensitive and less critical towards their peers. Amphetamines can be used to increase motivation of action. Neurosciences now seem to be able to explain, largely, the functions and biology of the brain, so that the neurotransmitters which are not doing their "job" correctly can be pointed out. Should one be depressed, they can receive antidepressants; should one be antisocial, they can rely on products which adjust the balance of neurotransmitters and reduce violent impulses. Could the "morality pill" strengthen self-control, empathy, goodwill, while also decreasing violent tendencies?

Besides neuropharmacological interventions, morality can also be enhanced by direct interventions on the brain with the help of noninvasive neurostimulation technologies. Transcranial magnetic stimulation (TMS), which is the less invasive, transmits magnetic impulses through the brain. A study conducted in 2006 concluded that this treatment can influence the way people respond to moral dilemmas. Transcranial direct current stimulation (tDCS) uses direct electrical currents to stimulate specific parts of the brain. It can increase cooperation and neuroplasticity, favouring the learning of pro-social behaviour [28]. Previous studies have shown that this treatment can influence the way in which people respond to incorrect offers and duties which imply moral judgement. Experimental neurostimulation has demonstrated that these treatments make people more involved in utilitarist judgements, from a moral point of view. Deep brain stimulation (DBS) is a neurosurgical procedure (which involves implanting electrodes in areas of the brain) which some believe can increase motivation [29], [30]. Cabrera et al. [31] highlights that using DBS, TMS and tDCS may produce headaches and can affect the personal identity in ways which cannot be anticipated (such as changing the thinking pattern).

J. Savulescu proposes a revolutionary technology for the purpose of moral enhancement, that is, genetic screening. Savulescu argues that when using in vitro fertilisation, parents have the moral obligation (not only liberty) to choose the embrio which can lead the life with the most well-being. The obligation is moral, not legal [32]. Genetic screening also implies the moral obligation of choosing the most ethical children: "Indeed, when it comes to screening out personality flaws, such as potential alcoholism, psychopathy and disposition violence, you could argue that people have a moral obligation to select ethically better children..." [33].

For the moment, human enhancement is the "hands" of pharmacology, but it will soon be linked to genetic engineering, nanotechnology, neurosciences, virtual reality, and other emerging fields, making its study increasingly challenging.

Moral Enhancement or Moral Treatment?

Recent debates focus on the difference between moral treatment and moral enhancement. Some philosophers argue that there is a difference between increasing the moral behaviour from an underdeveloped to a normally-developed level and from a normally-developed to a highly-developed level. Others define moral enhancement as any type of moral bettering, without using an average level to which moral abilities would be linked. N. Agar [34], for instance, considers that moral enhancement only refers to interventions which increase moral abilities to a level above the one already observed. N. Agar [35] also observed that therapeutical interventions can eventually modify deficiencies, while they can also "produce unbalanced excesses when used to enhance beyond human norms" (p. 369). Therefore, the two terms (moral enhancement and moral treatment) are different and should be used accordingly.

DeGrazia [36] defines enhancement as "any deliberate intervention that aims to improve an existing capacity, select for a desired capacity, or create a new capacity in a human being" (p. 1). This definition covers both biomedical and non-biomedical enhancement (such as education). Horskötter et al [37] argue that the interventions meant to restore normal moral abilities in subjects whose moral functioning was pathologically affected should be included in the sphere of moral treatment, not enhancement (p. 27).

The distinction between moral enhacement and treatment raises several issues. One of them is the normative level, as it implies the introduction of a level of normality between the insufficient moral functioning and the above-average moral functioning. Moral treatment would lie below this level of normality, wheareas moral enhancement would be above the level. There is no consent so far on where this average (normal) level of moral behaviour is situated. The question is: HOW does one determine the normal functioning limits of moral abilities? Which types of behaviours and disfunctionalities classify one person below the level of normality?

A distinction of the interventions used in both moral treatment and enhancement may also be necessary. However, there are cases when the same type of intervention (such as pharmaceutical) may be therapeutic at first, but once the person reaches the average level of morality, the intervention becomes enhancement. Therefore, when do we define intervention as a means of treatment or as a means of enhancement? Furthermore, there may be cases when an intervention with pharmaceutics has moral enhancement as side effect. Does that person want to be enhanced morally or is his/her choice violated? Should we consider that everything below the level of normality belongs to the treatment sphere, immorality may be justified as a medical issue. Who should determine whether one is normally moral? Or, when should normal functioning be diminished so as to enter the pathological sphere? Could emrbio genetic selection of the most ethical children be justified, as proposed by J. Savulescu? This work is not written from this perspective, but we consider that in time, as biotechnologies develop and the conceptual apparatus is better defined, the distinction between moral treatment and moral enhancement will bring new theoretical and practical clarifications.

Is moral enhancement possible, permitted, or mandatory?

As seen above, moral bioenhancement is possible through several means, at least on an individual level. For now, enhancement technologies are still in the early stages of development or they are still purely theoretical. Nevertheless, some pharmaceutics with medical prescription (drugs, pills), some supplements or stimulation devices (such as transcranial magnetic stimulation) are already used and available on the medical or black markets. Furthermore, some pharmaceutics with medical prescription, used on the wide scale, influence the moral behaviour of consumers. Neil Levy et al. [38] point to three drugs which modify moral decision and behaviour: propranolol, antidepressants (SSRI), and products which influence oxytocin. Similarly, a study conducted in 2013 by Terbeck et al. [38] concluded that subjects who use propranolol (for high blood pressure), men and women, are "more likely to judge sacrificial actions as morally unacceptable." Propranolol influenced morality especially in terms of personal dilemmas and seemed to increase aversion towards hurting others. Furthermore, "subjects who took propranolol were more likely to make deontological judgments, and to reject contrary utilitarian solutions to moral dilemmas, but only when these utilitarian solutions involved 'up close and personal' harm to an innocent victim" [38].

Antidepressants reduce aggression and may influence people into being friendlier or more sensitive towards others' pain. Pharmaceutics, like Ritalin and Adderal, improve the control over impulses in children with attention deficits, reducing violence. However, improving a characteristic involved in moral action may lead to diminishing another or even to immoral choices. For instance, increased self-control could lead to the lack of risk taking. Oxytocin was proved to enhance altruistic behaviour and empathy, but only within one's social group [39, p. 190]. Oxytocin could diminish cooperation with members outside the group, such as rasial minorities, promoting selective ethnocentrism, favouritism, and even aggression towards other groups [40]. Ruud ter Meulen highlighted that while some medicine can increase moral behaviour, others (or sometimes the very same) could have the opposite effect [41].

The preoccupation with this field, as well as the studies which showed that chemical substances in the brain can affect certain judgement and moral decisions, have stimulated imagination, thus discussions on the "morality pill" increased after the year 2011. Although a pill which could improve moral behaviour seems closer to fiction, Guy Kahane argues that although the field is only in its incipient phase, "it is very far from being science fiction" [42]. Kahane says that scientists will develop this type of medicine and will create new ways of ingesting them, giving the example of oxytocin taken as nasal spray [42]. Although we are yet to buy the "morality pill," the possibility of moral enhancement with the help of pharmaceutics generated debates whether such interventions should be permitted or if the obligation of moral bioenhancement should actually exist. The supporters of moral bioenhancement often give arguments on the benefits of strengthening pro-social behaviour, such as the inclination towards empathy, collaboration, aggression reduction, etc. The diminishment of acts of violence and of antisocial acts in general would imply a reduced number of incarcerated people. However, we highlighted before that strengthening some characteristics may lead to the diminishment of others and create imbalances which can generate immoral acts [43, p. 150]. Persson and Savulescu [44] [45] argue that moral enhancement should be mandatory. Savulescu argues for genetic screening, which may be a new form of eugenics [46]. He also states that "if safe moral enhancements are ever developed, there are strong reasons to believe that their use should be obligatory, like education or fluoride in the water, since those who should take them are least likely to be inclined to use them. That is, safe, effective moral enhancement would be compulsory."

Therefore, we consider that moral bioenhancement, on an individual level, will be possible, but within a timeframe which cannot be predicted. As seen, bioenhancement already exists, in certain incipient forms. Biotechnologies in the field of neurosciences and genetics is progressing so rapidly that what seemed impossible just a few decades ago, is now possible. The major interest for the study of the brain is reflected in the large budgets allocated to research projects so that while the 20th century was the century of the DNA, the 21st century may be regarded as the century of the brain. We mention within this context the Human Brain Project, with a budget of €1 billion over ten years (2013-2023), with support from the European Commission and from other partners including universities, industries and Member States' own programmes, and with the implication of over 400 researchers from 24 countries [47].

From individual moral enhancement to the moral enhancement of humanity?

Most definitions state that direct moral enhancement refers to interventions done on people individually. However, some scholars also include the interventions done on groups or even on humanity, in general. I. Persson and J. Savulescu discuss the improvement of the moral character of humanity. The enhanced characteristics which they discuss the most are altruism, trust in the members of the group, empathy, self-control (in the sense of reducing violence and anti-social behaviour), the will to cooperate, etc. As already obvious, the aim of these characteristics' enhancement is pro-social behaviour. However, definitions may differ with regard to the aim of moral enhancement. T. Douglas [48, p. 3] defines moral enhancement as the interventions which give a larger number of motivations and moral behaviours to an individual than (s)he would normally have. I. Persson and J. Savulescu argue for the enhancement of humanity through the moral obligation of selecting more moral children: "The genetically programmed child is in production ... People have a moral obligation to select ethically better children" [44] [45]. This implies to genetically examine embrios with the scope of determining which one has superior ethical qualities. What happens with the less ethical embrios? The moral enhancement of humanity would be based on a procedure involving genetic screening and selective embryo abortion. Such a program is contested and considered immoral by both some bioethicians and religious groups. The two authors nevertheless consider that methods should be put forward to enhance morality through pharmaceuticals, genetic editing, or any other method. The supporters of bioenhancement thus hope that new biotechnologies could contribute to a better world by improving moral reasoning, pro-social behaviour, and the strengthening of moral virtues. But would this be possible in reality? Would this be the path to a more ethical and moral world? And, furthermore, is morality an objective for all people?

The products and procedures mentioned above are debatable not only from a moral point of view, but also from the point of view of their production methods and their (side) effects. Although they seem to offer benefits to the society, they can have negative, unthought-of results. Because the ethical problems resulted from moral bioenhancement represent a large theme and are not the main purpose of this chapter, the following part will present a summary of some of these problematic aspects: favouring abuse; individual freedom; personal identity and undermining authonomy; lack of moral diversity and debates; the overestimation of moral enhancement; paying less attention to side effects, risks, and safety; a new eugenics by "projecting" moral children and embryonic selection; the risk of having, in time, the obligation of moral enhancement; the lack of freedom to act in an immoral way.

Why is bioenhancement not the path towards a more ethical world?

This section discusses the possibility of global moral bioenhancement starting from interventions such as the usage of pharmaceutics, neurostimulation, or genetic editing. Our premise is that moral bioenhancement, applied on the level of humanity, is not feasible. The interventions subtly influence pro-social behavior (empathy, lack of violence, voluntary action, etc.). However, morality is much more than this and cannot be reduced to its pro-social dimension, regardless of how important it is. Therefore, a broad definition of morality is needed, which would also encompass a set of universal and desirable virtues and traits capable of being enhanced. However, who should determine, globally, what is moral and immoral? This set of virtues and principles should be mandatory and accepted by everybody. We should also not neglect people's religious beliefs, which are different and solid. Therefore, before moral bioenhancement is implemented globally, a series of debates on global ethics should take place. Of course, bioethics had its path opened through Principlism, being a set of universally-applicable principles coming from collective morality. But moral bioenhancement raises specific ethical issues. If the project fails globally, two separate groups of people will exist: enhanced and non-enhanced, that is, moral people and the rest of them. It is known that there is no moral gene. However, researchers argue that biology (for instance, through the levels of oxytocin) can influence morality and external interventions would thus adjust/modify moral levels. Should those with low speficic biological values (such as oxytocin) be considered immoral? Which authority would determine the average level of these biological values beyond which a person would be considered moral? And how could such a procedure be controlled globally so as humanity would be enhanced? For the moment, this is an impossible endeavour. External interventions on neurotransmitters can lead to change in behaviour, but they do not change who the person is, they do not make that person better in his/her essence. Giving up on farmaceuticals, for instance, would most probably lead to the person going back to his/her original state. Who would have the global or even local control of such a process? Of course the benefits of a moral society are numerous, if its members voluntarily choose enhancement. They should be allowed to choose, not be manipulated or have paternal interventions inflicted on them. There will always be people who will not wish to be moral.

The solution of J. Savulescu and I. Persson is that of genetic screening, argued to be a moral obligation of parents who choose in vitro feritilisation. Would non-ethical embrios then be destroyed? Such a procedure is rejected by people with profound religious beliefs. Some issues still remain: that of taking responsibility, standardizing, and applying on global level. Even so, there will be children conceived naturally, born from unethical parents, who cannot, for the moment, be genetically screened.

Furthermore, there is the issue connected to the efficiency of these interventions, brought into attention by Veljko Dubljevic´ and Eric Racine [49]. They studied previous research done on using enhancement technologies, evaluating both their effects and the way in which they can be practically applied. Their conclusion is the following: "the predicted effects of state of the art of neurointervention simply do not materialize ... In sum, an analysis of current interventions leads to the conclusion that they are blunt 'instruments:' any enhancement effect is unspecific to the moral domain" [49, p. 348].

Conclusion

The field of neuroethics is not only an emerging one, but is also a challenging one as it has a tremendous dynamics, in the context of the development of other connecting fields which it intersects. Moral enhancement has a special place within debates, with some being interested in the enhancement of the individual, and others seeing it as way of making the world a better place. Athough some researchers argue for clear effects on human behaviour generated by neuropharmaceuticals, others argue in the complete different direction. We think that the possibilities of moral enhancement are somewhat overestimated which leads to exaggerations (at least for the moment), such as the enhancement of humanity. The risks of the interventions are numerous and the ethical problems remain unsolved. We do not exclude the possibility of moral enhancement (such as in the case of violence and not only), but it should be done through objective analyses, while trying to avoid excess on each side. Of course biotechnologies and pharmaceuticals can be an important resource for solving several moral problems, but only on an individual level, not on a community or global level. Once technology reaches a certain level, the advance in research cannot be stopped, but as much as moral enhancement progresses, we consider that it cannot replace traditional moral education. Education creates a change within the being, in its essence, and is accessible to everyone. Nevertheless, the "morality pill" may be closer to its achievement than we imagine, and if so, no boundaries can stop it.

References

[1] C.Elliott, *Better Than Well: American Medicine Meets the American Dream*. New York and London: W. W. Norton and Company, 2003.

[2] The President's Council on Bioethics, *Beyond Therapy: Biotechnology and the Pursuit of Happiness.* Washington, D.C., Oct. 2003, available at: https://bioethicsarchive.georgetown.edu/pcbe/reports/beyondtherapy/index.html. Accessed: April 15, 2017.

[3] A. Chatterjee, "Cosmetic neurology: the controversy over enhancing movement, mentation, and mood", *Neurology,* 63, 968–974, 2004. doi: 10.1212/01. WNL.0000138438.88589.7C

[4] B. D. Earp, A. Sandberg, G. Kahane, and J. Savulescu, "When is diminishment a form of enhancement? Rethinking the enhancement debate in biomedical ethics", *Frontiers in Systems Neuroscience*, 8 (12), pp. 1-8, 2014.

[5] I. Persson and J. Savulescu, *Unit for the Future. The Need for Moral Enhancement.* Oxford: Oxford University Press 2012.

[6] I. Persson and J. Savulescu. "Getting moral enhancement right: the desirability of moral bioenhancement". *Bioethics* 27, pp. 124–131, 2013. doi: 10.1111/j.1467-8519.2011.01907

[7] Merriam-Webster [Internet]. Springfield, MA: Merriam-Webster, Incorporated; 2015 [2015; August 1st, 2016], available at: http://www.merriam-webster.com/dictionary/morality. Accesed: April 10, 2017.

[8] B.Gert, *The Definition of Morality*.2016, available at: https://plato.stanford.edu/entries/morality-definition/ . Accessed: April 10, 2017.

[9] T.M. Scanlon, "Contractualism and Utilitarianism", *Utilitarianism and Beyond*, A. Sen and B. Williams eds., Cambridge: Cambridge University Press, pp. 103–28, 1982.

[10] S. Darwall, *The Second-person Standpoint: Morality, Respect, and Accountability*. Cambridge: Harvard University Press, 2005.

[11] W. Safire,"Introduction", *Neuroethics: Mapping the Field*, 1st edition, S. J. Marcus (ed.), New York: Dana Press, pp.3-10,2004.

[12] A. Roskies. "Neuroethics for the new millennium," *Neuron* 35, pp. 21-23, 2002.

[13] D.Thomas, "Moral Enhancement," *Enhancing Human Capacities*, edited by Julian Savulescu, Ruud ter Meulen, and Guy Kahane, eds., Wiley-Blackwell, Apr. 2011.

[14] K. Jebari, "What to Enhance: Behaviour, Emotion or Disposition?", *Neuroethics* 7(3), pp. 253-261, December 2014.

[15] I. Persson and J. Savulescu, "The perils of cognitive enhancement and the urgent imperative to enhance the moral character of humanity", *Journal of Applied Philosophy,* 25(3), pp.162–177, 2008.

[16] B. D. Earp and J.Savulescu, *Moral neuroenhancement*, July 2017, https://www.researchgate.net/publication/305776669 Accessed: Sept 10, 2017.

[17] *H. Wiseman, The Myth of the Moral Brain: The Limits of Moral Enhancement* . Cambridge, MA and London: The MIT Press, *2016*.

[18] P. S. Churchland, *Braintrust. What Neuroscience Tells Us about Morality*. Princeton, Princeton University Press, 2011.

[19] P. S. Churchland, *Touching a Nerve. The Self as Brain.* New York – London: W. W. Norton and Company, 2013.

[20] J. D. Greene, R. B. Sommerville, L. E. Nystrom, J. M. Darley and J. D. Cohen, "An fMRI investigation of emotional engagement in moral judgment", *Science,* 293 (5537), pp. 2105-2108, 2001.

[21] R. M. Bernhard, J. Chaponis, R. Siburian, P. Gallagher, K. Ransohoff, D. Wikler, R.H. Perlis and J. D. Greene, "Variation in the oxytocin receptor gene (OXTR) is associated with differences in moral judgment", *Social Cognitive and Affective Neuroscience*, pp. 1–10, 2016.

[22] J. F. Marques, N. Canessa, S. Cappa, "Neural differences in the processing of true and false sentences: insights into the nature of 'truth' in language comprehension", *Cortex* 45, pp. 759–768 [PubMed], 2009.

[23] V. Dubljević, E. Racine , "Moral Enhancement Meets Normative and Empirical Reality: Assessing the Practical Feasibility of Moral Enhancement Neurotechnologies", Bioethics, vol. 31, no. 5, pp. 338–348, 2017. doi:10.1111/bioe.12355

[24] D. G. Jones, "Moral Enhancement as a Technological Imperative", *Perspectives on Science and Christian Faith* 65, pp. 187–95, 2013.

[25] M. B. Regan, J. Chaponis, R. Siburian, P. Gallagher, K. Ransohoff, D. Wikler, R. H. Perlis and J. D. Greene, Variation in the oxytocin receptor gene (OXTR) is associated with differences in moral judgment", *Social Cognitive and Affective Neuroscience*, pp.1–10, 2016, doi: 10.1093/scan/nsw103

[26] T. Douglas, "Moral Enhancement", *Journal of Applied Philosophy,* 25, pp. 228–45, 2008.

[27] T.J. Trothen, "Moral Bioenhancement through An Intersectional Theo-Ethical Lens: Refocusing on Divine Image-Bearing and Interdependence", *Religions, 8* (5), 84, pp. 1-14, 2017. doi:10.3390/rel8050084.

[28] A. Piore, "A Shocking Way to Fix the Brain". *MIT Technology Review*, 2015, available at:
https://www. technologyreview.com/s/542176/a-shocking-way-to-fix-the-brain/ Accessed: April 20, 2017.

[29] V. Dubljevic, M. Shipman, "Study: 'Moral Enhancement' Technologies Are Neither Feasible Nor Wise".
https://news.ncsu.edu/2017/05/moral-enhancement-tech-2017/ Accessed: September 15, 2017.

[30] [23] V. Dubljević, E. Racine , "Moral Enhancement Meets Normative and Empirical Reality: Assessing the Practical Feasibility of Moral Enhancement Neurotechnologies", *Bioethics*, vol. 31, no. 5, pp. 338–348, 2017. doi:10.1111/bioe.12355

[31] L. Y. Cabrera, E. L. Evans and R. H. Hamilton, "Ethics of the ElectrifiedMind: Defining Issues and
Perspectives on the Principled Use of Brain Stimulation in Medical Research and Clinical Care", *Brain Topography* 27, pp. 33–45, 2014.

[32] J. Savulescu, '"Procreative beneficence: why we should select the best embryos," *Bioethics,* 15.5/6 pp. 413–426, 2001.

[33] *The Telegraph*, Genetically engineering 'ethical' babies is a moral obligation, Richard Alleyne , 16 Aug 2012, http://www.telegraph.co.uk/news/science/science-news/9480372/Genetically-engineering-ethical-babies-is-a-moral-obligation-says-Oxford-professor.html. Accessed: April 20, 2017.

[34] N. Agar, "Enhancing genetic virtue?", *Politics and the Life Sciences* 29(1), pp. 73–75, 2010. doi:10.2990/29_1_73.

[35] N. Agar, " A question about defining moral bioenhancement". *Journal of Medical Ethics*, 40(6), pp. 369–370, 2014.

[36] D. DeGrazia, "Moral enhancement, freedom, and what we (should) value in moral behaviour". *Journal of Medical Ethics*, pp.1-9, 2013. doi:10.1136/medethics-2012-101157.

[37] D. Horstkötter, R. Berghmans, G. de Wert, "Moral enhancement for antisocial behavior? An uneasy relationship", *AJOB Neurosci*, 3(4), pp. 26–28, 2012.

[38] N. Levy, T. Douglas, G. Kahane, S. Terbeck, P. J. Cowen, M. Hewstone and J.Savulescu, ,"Are You Morally Modified? The Moral Effects of Widely Used Pharmaceuticals", *Philos Psychiatr Psychol*, 21(2), pp.111–125, 2014. doi:10.1353/ppp.2014.0023.

[38] S. Terbeck, G.Kahane, S. McTavish, J. Savulescu, N. Levy, M. Hewstone, Philip J. Cowen, ,"Beta adrenergic blockade reduces utilitarian judgement", *Biological Psychology,* 92 (2), pp. 323–328, 2013.
doi: 10.1016/j.biopsycho.2012.09.005

[39] Jones, D. Gareth, "Moral Enhancement as a Technological Imperative", *Perspectives on Science and Christian Faith,* 65, pp. 187–95, 2013.

[40] C.K.W. De Dreu, L.L Greer, G.A. Van Kleef, S. Shalvi and M.J.J. Handgraaf, "Oxytocin promotes human ethnocentrism", *Proc Natl Acad Sci U S A.*; 108, pp. 1262–1266, 2011.

[41] [42] A. Hill, *Manipulating morals: scientists target drugs that improve behaviour,* available at:
https://www.theguardian.com/science/2011/apr/04/morality-drugs-improve-ethical-behaviour. Accessed: April 20, 2017.

[43] [24] D. G. Jones, "Moral Enhancement as a Technological Imperative", *Perspectives on Science and Christian Faith* 65, pp. 187–195, 2013.

[44] [6] I. Persson and J. Savulescu. "Getting moral enhancement right: the desirability of moral bioenhancement". *Bioethics* 27, pp. 124–131, 2013. doi: 10.1111/j.1467-8519.2011.01907

[45] I. Persson, and J. Savulescu, "The art of misunderstanding moral bioenhancement". *Camb. Q. Healthc. Ethics* 24, pp. 48–57, 2014. doi: 10.1017/S0963180114000292

[46] R. Alleyne, *Genetically engineering 'ethical' babies is a moral obligation,* Aug 2012 https://www.telegraph.co.uk/news/science/science-news/9480372/Genetically-engineering-ethical-babies-is-a-moral-obligation-says-Oxford-professor.html. Accessed: September 15, 2017.

[47] https://ec.europa.eu/digital-single-market/en/news/european-commission-signs-agreement-take-human-brain-project-hbp-next-phase. Accessed: September 20, 2017.

[48] T. Douglas, "Moral Enhancement Via Direct Emotion Modulation: A Reply to John
Harris", *Bioethics,* 27(3), pp. 160–168, 2011.
[49] [30] [23] V. Dubljević, E. Racine , "Moral Enhancement Meets Normative and Empirical Reality: Assessing the Practical Feasibility of Moral Enhancement Neurotechnologies", *Bioethics*, vol. 31, no. 5, pp. 338–348, 2017. doi:10.1111/bioe.12355

Public Deliberation and Biomedical Enhancements

Ivan Mladenović

Introduction

In this chapter, I will consider the significance of public deliberation for decision-making on bioethical issues, primarily those that concern biomedical enhancements. In the past couple of decades, deliberative democracy has become the dominant point of view in democratic theory. The essence of this approach is encapsulated in the provision that the process of decision-making on political issues should enable discussion and exchange of reasons among free and equal citizens. The basic idea is that through discussion and proffering of arguments in favour of opposing positions, an agreement on the best solution can be reached and consequently, better decisions can be made. If the first period of development of this theory was primarily preoccupied with the normative justification of the public deliberation procedure, today we are witnessing an increasing number of experiments with deliberative democracy and implementation of various kinds of public consultations on deliberative grounds.

There is no doubt that bioethics is the field marked by numerous reasonable disagreements. However, some authors have expressed a view that it is precisely because of this that deliberative democracy can be of immense importance in this context [1]. Whether citizens themselves make decisions regarding bioethical issues or whether institutions do it, deliberative democrats think that "accountability for reasonableness" is crucially important to the whole process [2]. It means that regarding controversial bioethical decisions, either citizens among themselves or institutions, should furnish reasons for their decisions that should be acceptable to all. Even if the requirement of reasonableness in this strong sense may be just a normative ideal, what seems to be practically feasible is to demand that a procedure of public deliberation be held regarding controversial bioethical decisions, a procedure that will enable justifying certain decisions, as well as revision of these decisions on reasonable grounds. Even if reasonable disagreements persist after discussion has been conducted, public deliberation will nevertheless have contributed to legitimacy and fairness because it is based on an aspiration to offer justification for a certain decision. Similarly to developments within deliberative democratic theory, from the first stage which was mostly characterized by normative considerations, the discussion has largely moved towards practice of public deliberation regarding various bioethical issues [3–5].

However, in this chapter I will primarily discuss the role of public deliberation on human biomedical enhancements. This is the reason why I will discuss the view of Alan Buchanan, who maintains that in order to face problems associated with biomedical enhancements, it is crucially important that there be an appropriate institutional framework [6]. Namely, according to Buchanan, the best way to handle this issue is the enhancement enterprise. A significant part of the enhancement enterprise is public deliberation on moral justification of biomedical enhancements as well as the risks associated with undertaking them. Furthermore, the essential importance of public deliberation inheres in the possibility of revision of previously-made decisions in the light of new evidence. If public deliberation is to function at all, wide accessibility of information relevant to the issue of biomedical enhancements would be necessary. Although I agree with Buchanan that orientation towards institutions and public deliberation is justified, I will criticize his point of view as insufficiently precise in terms of how the very procedure of public deliberation should look like. In order to additionally specify the issues regarding public deliberation on biomedical enhancements, I will introduce the difference between micro-deliberation and macro-deliberation. Taking into account the literature regarding experiments with public deliberation, I will point out in what way public deliberation on biomedical enhancements should be organized at the micro and at the macro level.

Deliberative Democracy and Bioethics

The dominant stand in democratic theory is that democracy should function on deliberative grounds. Electoral democracy does not leave enough room for citizens to voice their reasons in public discussion and to bring informed decisions on the basis of a better argument. Therefore deliberative democracy theorists maintain that the process of democratic decision-making should be deliberative to a greater extent.

Consequently, it could be said that public deliberation plays several important roles. First, it contributes to the legitimacy of democratic decision-making. Facilitated by public deliberation, democratic decision-making significantly approximates an ideal that all individuals should be treated as free and equal, but also as reasonable and rational citizens. Second, it has an epistemic role because it contributes to better decisions which are based on reasons that have passed the test of public re-examination. The basic idea is that better decisions will be brought if they are previously discussed in view of a better argument. Third, it contributes to fairness of decision-making, because even those whose point of view has not been adopted can see that in the very process of decision-making they have been treated fairly. Moreover, public deliberation promotes thinking in terms of the common good, given that it is based on public reasons that should also be acceptable to other citizens. On the basis of these characteristics of public deliberation, some theorists have concluded that it should be aimed at arriving at a consensus about the common good. However, it cannot be said that there is any real necessity in this regard. Even if process of public deliberation should lead to reasonable disagreement, it can still be claimed that decisions brought by majority voting that had been preceded by this form of discussion are justified because the very procedure contributes to legitimacy, quality and fairness of decision-making, even if not everybody agree with these decisions.

Gutmann and Thompson have contributed significantly to the theory of deliberative democracy. But they were also among the first authors to notice the relevance of deliberative democracy for bioethical issues. Namely, Gutmann and Thompson maintain that the following four characteristics of deliberative democracy can contribute to making better decisions about bioethical issues that are the subject of moral disagreement: legitimacy of collective decision-making, promoting public-oriented perspectives, mutual recognition when decisions are brought and possibility of correcting mistakes [1]. The legitimacy of decision-making is important because bioethical controversies often pertain to disagreements regarding scarce resources. In the situation in which not everyone can acquire certain resources, it is extremely important to justify priorities by furnishing reasons that could be acceptable to all. Given that limited altruism is a frequent source of moral disagreements, public deliberation can contribute to taking an impartial perspective. It particularly concerns bioethical commissions and other similar bodies in which participants should not represent certain group interests, but should take decisions from an impartial point of view. Mutual recognition is important for decision-making because even in situations of disagreement on some fundamental values, there is often room for agreement regarding many other issues. It is precisely mutual recognition during public deliberation that can help narrow the space for disagreements that are often a result of lack of information or some other factor that can be corrected in a discussion. Finally, public deliberation enables us to point out possible errors in judgment, but also to correct wrong decisions which had previously been taken.

A similar view is propounded by Norman Daniels who particularly stresses the role of public deliberation for solving bioethical issues, primarily those that pertain to defining priorities regarding health care needs. Daniels maintains that "accountability for reasonableness" is a crucial aspect for ensuring legitimacy and fairness of decision-making on such issues. "Accountability for reasonableness" means that institutions where decisions are made on issues that are the subject of moral disagreement should function in accordance to several conditions characteristic to deliberative democracy [2]. According to this view, even if in the very process of decision-making those affected by decisions do not take part, they can be legitimate and fair if the process of decision-making is transparent i.e. there is a public justification for the given decision. Secondly, this justification has to be reasonable in the sense that it is based on reasons that are acceptable to all. And thirdly, this process enables to raise an objection to the given decision in order to revise it if there are justified reasons to do so. To the aforementioned conditions, sometimes a fourth condition is added that concerns the necessity for either voluntary or state regulation for ensuring the previous three conditions. What is important for the ensuing discussion is that these conditions should also hold for fair redistribution of new treatment technologies, which may be relevant regarding new technologies for biomedical enhancements. In this chapter, I will consider precisely the importance of public deliberation for issues of biomedical enhancements.

The Enhancement Enterprise and Public Deliberation

The first stage of the discussion on biomedical enhancement mostly concerned giving reasons against or for enhancement as such. Despite the fact that this dialogue is far from over, it could be said that in the meantime the discussion assumed yet another level. This topical stage of discussion primarily concerns the following questions: which enhancement and which means of human enhancement would be morally permissible if we assume that some enhancements are necessary to ensure a meaningful life on earth [7]. The main argument of those arguing against enhancement is based on the premise that radical intervention in human nature which would be made in order to enhance it, could lead to its full annihilation. A typical attitude of proponents of enhancement, could be that enhancement as such is something good. But, on the other hand, she may also argue that if not all, then at least some types of enhancement can be morally permissible. For example, certain types of enhancement may be necessary in order to ensure the survival of people on planet Earth which could be significantly jeopardized by climate change. However, a question is raised: which things ought to be subject to change, in what way and to what extent? And at this point, the views, even among proponents, of enhancement differ.

Alan Buchanan thinks that rather than discussion devoted to purely theoretical arguments for and against enhancements, it would be much more fruitful to accept an enhancement enterprise that encompasses institutions within which such issues could be resolved. This is why public deliberation, in his view, is an integral part of the enhancement enterprise. Buchanan claims that the idea about enhancement enterprises actually suggests that certain biomedical enhancements can be legitimate goals for individuals and institutions. His main argument in favour of the enhancement enterprise is the following. He proposes a sort of a hypothetical choice situation. The first option is that the issue of biomedical enhancements should remain completely unregulated. In that case, tools for combating negative effects of certain enhancements on the social level would also be lacking. The absence of any regulation can lead to unfair distribution of medical enhancements; it can unfold utterly non-transparently and without sufficient knowledge about the risks certain biomedical enhancements involve. The second option is that the issue of biomedical enhancements should be regulated and that the justification of certain enhancements should be the subject of discussion within appropriate institutions. Such an institutional perspective would allow to adequately solve problems regarding the fair distribution of biomedical enhancements. Furthermore, another vital element should be public deliberation on the basis of which arguments would be put forward for and against certain enhancements. Buchanan maintains that in a hypothetical choice situation, the second option which favours the enhancement enterprise should be chosen. He also claims that his standpoint should not be interpreted to claim that biomedical enhancements as such are justified. Namely, in the process of public deliberation, some biomedical enhancements could be rejected on the basis of justified reasons. Thus according to Buchanan, institutional perspective would offer a way out of the deadlock in which purely theoretical arguing for and against enhancements would lead.

According to Buchanan, public deliberation could play several important roles within the enhancement enterprise. On the basis of the previous argument, Buchanan concludes that enhancements should be "a legitimate topic for democratic deliberation, rather than allowing enhancements to come through the backdoor, unscrutinized and unregulated" [6, p. 73]. He also points out that biomedical enhancements are "a sufficiently controversial and risky undertaking to warrant ongoing public deliberation and institutional control" [6, p. 151]. In order to ponder the risks of biomedical enhancements, it is necessary to have a broad availability of the information and public fora where wrong decisions could be revised. It is important to point out that Buchanan does not think that the enhancement enterprise could be justified in any institutional context. Namely, he particularly emphasizes that such an enterprise presupposes a constitutional democracy, which already has appropriate mechanisms such as fair elections, checks and balances, protection of individual rights, for an enhancement enterprise to be able to be subjected to democratic control [6, p. 61].

However, although I agree with Buchanan's idea about the importance of institutions and public deliberation for tackling questions of biomedical enhancements, I think that there are several objections which can be made regarding this standpoint. First, by arguing in favour of the enhancement enterprise, Buchanan seems to assume what has to be proved. Namely, he assumes the legitimacy of the enhancement enterprise, which suggests that biomedical enhancements can be legitimate individual and social goals. However, if one insists on the importance of deliberative democracy, then it is not clear why the legitimacy of the enhancement enterprise should not also be subject to public scrutiny. Obviously, in that case the very enhancement enterprise could be shown to be insufficiently justified. The essence is that one cannot assume in advance what should only be the result of public deliberation; consequently, the issue of legitimacy of the enhancement enterprise would have to remain open until a decision on it is brought in a public forum. The second objection is directly related to the first, but does not pertain to the justification of the enhancement enterprise but to the justification of certain biomedical enhancements. As we have seen, Buchanan claims that on the basis of public deliberation, some enhancements could be rejected as unjustified. However, he offers a series of arguments in favour of numerous enhancements that include cognitive enhancements for increasing productivity, enhancements that contribute to the extension of life expectancy, enhancements of the immune system, moral enhancements, etc, as legitimate goals for individuals and institutions [6]. This contradicts his thesis that the justification or rejection of certain enhancements should ensue from the process of public deliberation. Therefore, instead of assuming in advance that certain biomedical enhancements are justified, one should wait for the results of public deliberation. The process of public deliberation might show that the biomedical enhancements which Buchanan finds justified are indeed acceptable, while it could reject some of them. In any case, a decision cannot be made prior to the very process of public deliberation, at least if we assume that the decision on certain enhancements should be brought in a public forum.

It seems that the additional problem is related to the insufficient specification of the process of public deliberation. Some of the problems we have previously pointed out could be overcome by additionally specifying the procedure of public deliberation. Namely, within an empirically-oriented deliberative democracy research, differentiation has been made between micro-deliberation and macro-deliberation [8]. Experiments regarding micro-deliberation such as deliberative polls have shown that deliberative democracy can successfully function in smaller groups. However, the problem is how to organize public deliberation of a wide scope. On the basis of experiments with deliberative polls, some theorists have concluded that with the assistance of the simultaneous functioning of a multitude of such fora, it is possible to conduct macro-deliberations on the level of an entire society. Bearing this in mind, it could be said that questions such as the justifiability of the enhancement enterprise should be subject of macro-deliberation that would include the largest possible number of citizens. The discussions that would concern justifiability of particular enhancements could function according to the model of micro-deliberation, which includes some fora in which only citizens take part, some in which only experts take part, as well as those in which both citizens and experts take part. However, even though these micro-deliberative fora could in some cases furnish reasons that are acceptable to all although not everyone can take part in them, the legitimacy of the enhancement enterprise should ensue from the broad participation of citizens at the macro-deliberative level.

Still, some theorists remain sceptical about the role of public deliberation regarding bioethical issues and particularly issues regarding biomedical enhancements. Namely, in a recent article, Clarke and Roache have expounded arguments that libertarian paternalism should be given priority over deliberative democracy when transformative technologies are concerned, including means for biomedical enhancements [9]. Their main argument is that public deliberation, instead of a consensus, is more likely to lead to further disagreements among citizens regarding these issues. An important premise in their argument is that ideological differences which are based on intuitions would only be augmented during public deliberation. Conversely, libertarian paternalism, with its technique of "choice architecture," is much more likely to lead to consensus and agreement regarding transformative technologies. However, even though in initial formulations of deliberative democratic theory, the significance of a consensus has indeed been strongly emphasized, in later elaborations of this theory, the view has prevailed that it is not necessary that public deliberation should lead to unanimous agreement. What is important is that participants in discussion proffer public reasons in favour of their point of view. Moreover, the procedure of public deliberation, rather than bolstering positions based on intuitions, may lead to exposing prejudices and acceptance of attitudes on rational grounds. The essence of the deliberative approach is precisely that although not everyone will agree with the final decision, everyone will be able to see that in the process of public deliberation, one has been treated fairly.

Conclusion

Some biomedical enhancements are already available today and it is quite certain that a larger number of such enhancements will be available in the future. In this chapter, I have considered the importance of public deliberation for facing the challenges presented by biomedical enhancements. I emphasized the role of public deliberation in the context of bioethics and subsequently for institutional handling of challenges posed by biomedical enhancements. In this context, I have particularly considered Buchanan's justification of the enhancement enterprise. I agree with Buchanan's view on the importance of institutions and public deliberation in this context. However, I think that the legitimacy of the enhancement enterprise should not be assumed in advance, but should itself be the subject of public deliberation.

References

[1] A. Gutmann and D. Thompson, "Deliberating about Bioethics," *The Hastings Center Report*, vol. 27, no. 3, pp. 38-41, May - Jun. 1997.

[2] N. Daniels, "Accountability for Reasonableness," *British Medical Journal*, vol. 321, pp. 1300-1301, Nov. 2000.

[3] J. Abelson, E. Blacksher, S. Boesveld, S. D. Goold and K. Li, "Public Deliberation in Health Policy and Bioethics: Mapping an Emerging, Interdisciplinary Field," *Journal of Public Deliberation*, vol. 9, no. 1, pp. 1-35, Apr. 2013.

[4] J. Lander, T. Hainz, I. Hirschberg, S. Bossert and D. Strech, "Do Public Involvement Activities in Biomedical Research and Innovation Recruit Representatively? A Systematic Qualitative Review," *Public Health Genomics*, vol. 19, no. 4, pp. 193-202, Aug. 2016.

[5] A. Moore, "Public Bioethics and Deliberative Democracy," *Political Studies*, vol. 58, no. 4, pp. 715-730, Oct. 2010.

[6] A. Buchanan, *Beyond Humanity? The Ethics of Biomedical Enhancement.* Oxford: Oxford University Press, 2011.

[7] I. Persson and J. Savulescu, *Unfit for the Future: The Need for Moral Enhancement.* Oxford: Oxford University Press, 2012.

[8] J. S. Fishkin, *When the People Speak: Deliberative Democracy and Public Consultation.* Oxford: Oxford University Press, 2009.

[9] S. Clarke and R. Roache, "Introducing Transformative Technologies into Democratic Societies," *Philosophy & Technology*, vol. 25, no. 1, pp. 27-45, March 2012.

PART III
SPECIFIC MEDICAL ETHICAL ISSUES

On the Ethical Issues of Bilateral and Contralateral Risk-Reducing Mastectomy

Gergely Tari, Gábor Braunitzer

Introduction

In traditional medical ethics, the central values generally supported the patients' medical interests, *sensu stricto*. In the Hippocratic tradition, the relationship between the physician and the patient is hierarchical, and their communication is unidirectional, as the goal is to "fix" the patient's *bodily* ailments from the external perspective of benevolent paternalism. However, this model is superseded - in theory at least. In their seminal work, *Principles of Biomedical Ethics* [1], Beauchamp and Childress argue that the guiding principle of medical ethics should be *the patient's autonomy*. That is, the silent and passive patient of the Hippocratic model should turn into an active participant and agent, a partner in *Shared Decision Making* [2]. Regardless of this theoretical development, it is still often questioned whether patients should freely decide on their therapy and in what cases should one's autonomy subordinate to benevolent paternalism.

The principles of *beneficence* and *patient autonomy* can clash in various ways; this is the case of bilateral risk-reducing (BRRM) and contralateral risk-reducing mastectomy (CRRM). Mastectomy as a therapeutic option, however, is rarely treated as an ethical issue. The literature of BRRM and CRRM is dominated by the topic of risk-assessment and the identification of high-risk patient groups.

This chapter will analyze BRRM and CRRM from a moral point of view, applying Beauchamp and Childress's principles. We are going to argue that *patients' autonomy* and women's self-determination over their own bodies can be considered as an indication of risk-reducing mastectomies. To support this, we examine other medical interventions that result in quasi-irreversible outcomes, such as contraceptive sterilization and the refusal of life-saving and life-sustaining interventions. Finally, we introduce an ethical guideline meant to complete the medical assessment criteria.

The history of prophylactic risk-reducing mastectomy

In medical literature, only a few cases from the 19th century on report surgeries where surgeons had removed "benign breast masses," and the first prophylactic mastectomy was carried out in 1917 [3]. By the 1970s, almost a thousand prophylactic mastectomies were being done annually in the United States [4]. This rise sparked some professional debate, but this debate was more medical than ethical in nature. Questions arose regarding the aim of BRRM and CRRM (fear management or actual cancer prevention) and the necessary extent of surgeries [3].

The picture has considerably changed over the past 40 years. According to recent statistics, the number of BRRMs and CRRMs tripled between 2002 and 2012, despite that no superior long-term survival has been confirmed as compared to the conservative therapies. Wong et al. examined 496,488 women diagnosed with unilateral invasive breast cancer. 59.6% chose breast-conserving surgery, 33.4% underwent unilateral mastectomy, and CRRM was chosen by 7.0%. All in all, the number of women who opted for CRRM increased from 3.9% in 2002 to 12.7% in 2012 [5].

Women's interest in risk-reducing mastectomy has globally increased. The media has a remarkable role in this. Celebrities' effect on BRRM, CRRM, and genetic screening for BRCA1/BRCA2 mutations has already been examined. The most attention was gained by Angelina Jolie's bilateral risk-reducing mastectomy in May 2013 [6-7]. After the AJ case, the number of BRRMs and the demand for BRCA-testing measurably increased. However, Evans et al. notice that there was no increase in 'inappropriate referrals.' [8]

Normative implications of BRRM and CRRM

Distinguishing between appropriate and inappropriate referrals generally creates a normative frame as well. The moral nature of these types of surgeries is basically related to two ethical issues. (1) If yes, by which limitations can medicine do quasi-irreversible harm to patients upon their request? *Autonomy* is the essence of the modern doctor-patient relationship, as well as *informed consent.* However, it is questioned whether detailed medical information – even repeated over several consultations – could be enough as a moral counterweight in case of removing an otherwise cancer-free breast. This is the first question we deal with. (2) Second, we examine those cases in which paternalism in the doctor-patient relationship can be morally justified. We will analyze whether patients' autonomy and self-determination can be reasonably limited regarding CRRM and BRRM. There is a widely supported moral guideline that states that one's freedom ends where another one's begins. In cases in which one does not put another person at risk, limiting the formers' self-determination could lead to unnecessary paternalism. On the next few pages, we argue that *benevolent paternalism* that exclusively aims to maximize patients' medical welfare is not justifiable when it comes to CRRM and BRRM requests. Instead of limiting the access to CRRM and BRRM, more emphasis should be put on effective informed consent in the framework of Shared Decision Making (SDM).

Informed consent and risk-reducing mastectomy

Until about the middle of the 20th century, the hierarchical nature of the paternalistic physician-patient relationship limited the information flow between the two parties, wherefore consent to any treatment was usually not sufficiently informed or rooted in real understanding. It comes as no surprise that Katz called the history of physician-patient relationship 'the history of silence' [9]. The normative turning point in the doctor-patient relationship changed the basic ethical standards regarding therapeutic decision making. According to the modern bioethical standards, patients' consent must be informed and the therapy has to be chosen with the patient's active involvement in the decision-making process. Chewning et al. analyzed 119 studies that focused on the attitudes of patients toward being involved in the making of such decisions. They found that 63% (n=75) of the studies concluded that patients want to be involved [10]. However, several meta-analyses emphasize that the desire of patients to be an active member of the decision-making process is not general [11-13]. To assume that all patients would like to have complex and detailed medical information and participate in the decision- making process is inaccurate. However, it does not mean that those patients who refuse to listen to the relevant medical information will benefit all the same from the therapeutic process. Patients who less actively cooperate with their doctors are less satisfied with their medical care [14]. All in all, the comprehensive information of patients and their involvement in the decision making process is essential, but it is just as important to respect their right to refuse to listen to the medical information to abstain from the making of therapeutic decisions. We argue that reference to these facts should be part of the basic information that the physician provides, preferably during the first encounter with the patient. That is, the patient should be aware that while participation in medical decision-making is not obligatory, those who get involved in that way turn out to be more satisfied.

Patient autonomy and breast cancer

Based on the above, it is obvious that providing autonomy in cancer has undeniable benefits, such as greater satisfaction and a better sense of control. The prerequisite of autonomy is that the physician and the patient participate in the interaction as partners. Emmanuel et al. distinguish between four basic models of the doctor-patient relationship [15]. The traditional and most ancient form is the so-called *paternalistic model,* in which all the relevant therapeutic decisions are made solely by the physician himself. The second type is the *informative model*, in which the doctor informs the patient, but still decides alone. The third model is much closer to the modern medical practice, because in the *interpretive model,* physicians explore their patients' values and inform based on those perceived values. The shared problem with these models is that the patient remains in the shadow of the doctor, therefore has no influence whatsoever on the course of the therapy. This way, autonomy is suppressed. Emmanuel and his colleagues ultimately argue that the optimal relationship between the two parties must be equal in which comprehensive and personalized medical information is provided, and then the doctor and patient make the decisions together. This desirable model is called the *deliberative model* and is in accordance with the modern bioethical standards. In a medical system that applies this model, the right to self-determination and the right to freely influence one's medical care is widely supported. *In summary, regarding breast cancer treatments, it is essential to provide all the patients personalized medical information as well as to provide them the chance to give informed consent to the patient-preferred medical treatment.* As we saw before, cooperation and informed consent improves a range of parameters from satisfaction to the sense of control. There is a number of other reasons for which the paternalistic model just cannot work in breast cancer cases. Each therapeutic decision is influenced by the decision makers' original life situation, value system, goals and fears, etc. The diagnosis of breast cancer is challenging for both the patient and the physician because of this moral/ethical complexity. Patients' attitudes towards the therapies are influenced by the access to health care, the fear of recurrence or occurrence, as well as the effect of a surgical intervention on one's personal appearance (body image) and sexual life [16]. The physician just cannot evaluate these parameters for the individual patient, wherefore all possible efforts must be made to engage patients in decisions about themselves.

The partner-like relationship optimally establishes the ground for the patient-centered Shared Decision Making (SDM) process in the oncological care as well, so that evidence- based information can be synchronized with the patient's relative preferences. According to Martinez et al. [15, p. 55], the most important aim of SDM is to support the patient to make a medically sound decision that is at the same time compatible with her value system. The physician has two roles in this process. (1) First, he informs the patient about every relevant detail regarding the treatment options, including their advantages and disadvantages in a comprehensible manner; (2) Second, the physician directly asks for his patients preferences regarding each alternative. It is essential to emphasize that SDM is not complete if the practitioner only lists the treatment options to his patient expecting the patient to decide on his own.

Shared decision making is a dynamic, two-sided process. In the lack of medical knowledge, we must not expect patients to be able to choose between the alternatives. SDM's primarily aims are to maximize patient welfare with the active assistance of the doctors' expertise exploring his patient's preferences towards a treatment's advantages, side effects, short- and long-term consequences.

Personalized medical information is crucial in the era of informed consent. Supporting patients toward an informed and deliberate consent is substantial. It can be facilitated with accessible decision supporting tools, as emphasized in several studies [17]. Waljee et al., in their randomized control study found that decision aids significantly (p=.024) improved patients' therapy-related knowledge. Additionally, these tools are suitable to eliminate treatment-decision related conflicts and improve satisfaction rates concerning decision making. *Hence, using such decision supporting tools like online interactive information materials or leaflets should be an integrative part of SDM* [18].

Patient autonomy and informed consent in relation to mastectomy

There are many studies focusing on therapeutic decision making of people who are diagnosed with breast cancer. This paper cannot analyze all the treatment options for breast cancer, thus we restrict our focus on the surgical decisions, or more specifically on the normative concerns of bilateral risk-reducing (BRRM) and contralateral risk-reducing mastectomy (CRRM). However, we also have to talk about an otherwise broadly accepted alternative to mastectomy, the breast conserving surgery (BCS). It is indispensable, because mastectomy and BCS can both be treatment options for many women who are diagnosed with breast cancer. The moral problem comes from this, because the range of potential complications is much wider for mastectomy than for BCS, while their effect on long term survival is not significantly different [15, p. 57]. In the light of these medical facts, it is almost imperative to search for the reasons why an increasing number of breast cancer patients opt for mastectomy of some kind. As long as the effect on long- term survival rates of BCS and mastectomy are in balance, should doctors allow their patients to choose a more invasive and extensive surgery?

The increase in CRRM and BRRM cases is presumably linked to increased access, which is not necessarily accompanied by in-depth understanding. Mastectomy and breast cancer are matters of public discussion/debate much more than they used to be before the millennium. At the same time, the patient population in question is not really aware, or does not actually trust the medical evidence regarding the similar survival rates of the surgical alternatives. This could be professionally and at the same time morally challenging to the medical team [19].

It is a well-known fact, that a huge proportion of women who are diagnosed with an early stage of breast cancer choose CRRM instead of BCS [20]. *This raises the suspicion that the doctrine of informed consent is violated to different degrees, since it is difficult see why women would choose a more invasive and extensive type of surgery in possession of all available information.* Even if there is reliable medical evidence that contralateral cancer recurrence is rare, many patients wish to undergo mastectomy because of the fear of recurrence [21]. *It seems that even though the patients are generally more conscious and better informed, they do not necessarily favor the medically more supportable options.*

Limitations of patient autonomy

If BCS is chosen by the patient, then the *nil nocere* (do no harm) principle also prevails, as the goal is achieved by a less invasive method. This is the optimal case. But let us imagine a bit more complicated scenario. It is conceivable that weighing the pros and cons of each treatment option, the surgeon would prefer BCS, but the patient holds on to mastectomy. If the doctor decides to suppress his patient's wish, it is a breach of autonomy. Women diagnosed with breast cancer, after having received comprehensive medical information, can still choose a treatment option that is medically not the "best," one that the surgeon cannot identify with. *The progress of medicine cannot be measured solely by focusing on the decreasing mortality rates and longevity, because nowadays we need to complete the former approach with the patient's evaluation of each treatment's outcome, risks, etc* [19, p. 5]. The origo of all breast cancer treatments is the patient herself, whose right to self-determination and autonomy must be protected. In the above thought experiment, the imaginary surgeon would impose unnecessary paternalism. It is a moral obligation of a physician who respects patients' autonomy to inform them about all the relevant details in a comprehensive and objective manner. Regarding CRM and BRRM, the additional disadvantages must be shared as well. All in all, only the patient herself is entitled to evaluate whether to accept the higher chance of sepsis, infections, pain or even rare mortality [22].

Everyone could harm themselves

In the field of medicine, limitation of patient autonomy is accepted only under certain circumstances. These circumstances are well defined and special. Cases of public health emergency, life-threatening emergency situations, incompetent (minor, mentally ill, etc.) patients are often cited cases regarding the limitation of informed consent [23]. These are situations when the actual risk overrules the relatively abstract concept of autonomy. There exist controversial treatment options that are seemingly clearly harmful or too risky from the medical point of view. In case of disputable patient requests, doctors tend to back out, in order to save patients from "irrational decisions." Nevertheless, it is questionable whether a request that is illogical or irrational from the physician's point of view, grants the right to the physician to curb patient autonomy. *Suppressing an otherwise competent patient's request on the grounds that it is not compatible with the physician's value system would lead to unjustifiable paternalism, even if the physician-preferred treatment option would promote the wellbeing of the patient. Accepting or rejecting a treatment decision of a patient should not solely depend on its' rationality, but rather on the patient's capacity and informed consent.* Julien Savulescu argues:

> *When doctors or others disagree with people's values or probability estimates, they should reason with them and engage them in normative dialogue. But if the patient is competent, the best reason for not respecting their choices is not the choice is imprudent or irrational, but on the basis of justice considerations and the fair allocation of medical resources, or on the basis of harm to others. The importance of freedom to construct our own conception of the good life, and to act on it, requires that doctors respect irrational choices, and, where resources allow, facilitate the originality and diversity of human existence.* [24, p. 34-35]

Examples to voluntary self-harm

In support of BRRM and CRRM it is worth considering two commonly supported medical therapies that result in physical harm upon the voluntary request of a competent individual. This is to show that risk-reducing mastectomy rises less serious ethical dilemmas than usually thought. That is, it often happens in the field of medicine that the two basic ethical principles namely "nil nocere" and patient autonomy clash and it does not often lead to unresolvable conflicts.

In the major part of the world, contraceptive sterilization of men and women became the most widely used method for couples in the reproductive age [25]. Sterilization as a permanent method for family planning is, however, still the subject of bioethical debates, since it is a quasi-irreversible contraceptive alternative. The ethical issue from the medical point of view stems from the conflict of the basic bioethical principles referred to above. The moral dilemma has been solved with different legal solutions in different countries, but nowadays the number of states that strictly restrict the use of this highly effective intervention is decreasing. This is appears to be a clear sign that the world tends to opt for patient autonomy in this respect. According to the International Federation of Gynecology and Obstetrics (FIGO) comprehensive and understandable medical information is the only moral counterweight that is needed. Any further limitation, such as the consideration of parity, age, waiting time, etc. would unreasonably restrict patient autonomy [26].

Another remarkable example of the same dilemma is the right to refuse medical care, even if it is life-saving or life-sustaining. Similarly to contraceptive sterilization, the ways and means of how one can practice the right of refusal depends on the legal environment of the given country. In general, in spite of the genuine moral difficulty, patients' therapeutic freedom overrules the medically beneficial decisions, therefore people have the ultimate right to accept or not a medical treatment. Of course, moral counterweights are needed here as well. Informed consent is a universally supported tool as the minimum ethical standard, and additionally several different are in use, such as waiting time [27].

Through these examples we tried to show that medicine does allow patients to choose treatment options that are irrational or at least controversial at first sight. However, such ethically loaded practices necessitate transparent guidelines and protocols. Below we give a brief description of a guideline related to such issues, which we find an ideal blend of medical and ethical considerations.

Guideline to risk-reducing mastectomy

The literature of BRRM and CRRM are mainly limited to the risk-assessment and almost no moral or ethical issues are discussed. Another frustrating gap is that most of the studies focus on the higher-risk group patients. Consequently, the therapeutic preferences of lower-risk patient are barely discussed. The *Manchester guidelines for contralateral risk-reducing mastectomy* however is a rare exception [28]. The discussion of CRRM is interweaved by all the necessary ethical standards, and accordingly the preoperative stage of contralateral mastectomy is under the influence of SDM. In practice, it seems quite easy; the patient and her physician consult multiple times before she finally gives her consent to the chosen treatment. The Manchester guidelines suggests a five-step evaluation of each patient who request CRRM. It starts with *history taking.* Open-ended questions are emphasized, hence doctors can understand the motives underlying patient decisions. The second step aims to *calculate the risk of contralateral breast cancer.* As a result, patients are classified to one of four risk-groups. The third stage of the evaluation process recommends a well-known ethical counterweight, namely waiting time. *The cooling off period* aims to reduce the number of bad decisions that may stem from the emotionally vulnerable period after the emotional vulnerability in the period after being informed of the diagnosis. Reviewing the literature of post-mastectomy regret, it is proven that 90% of women who chose risk-reducing mastectomy are satisfied with their decision for a longer term [29-30].

However, we argue that if medically agreeable, it is worth considering a shorter or longer waiting time before performing any type of risk-reducing mastectomy, as well as multiple doctor-patient consultations. The fourth step of the Manchester guidelines is the assessment of each CRRM request in a *multi-disciplinary (MDT) setting.* The team includes "breast cancer nurse, breast surgeon, oncologist, radiologist, pathologist and where possible, an oncoplastic-reconstructive surgeon familiar with the various reconstructive options including free TRAM/DIEP." [28, p. 4] Psychological support is a part of the MDT as well. The team estimates the risk of the contralateral breast cancer, assesses the chance of survival, outcomes and side-effects of each treatment option. Since the comprehensible and personalized medical information is essential to facilitate SDM, the MDT could enhance the application theory to practice. Finally, before the patient makes her informed consent the physician summarizes all the possible outcomes, side effects and long term consequences of each treatment option. It is only after this that the patient can sign the informed consent form.

Conclusion

Choosing the appropriate treatment option for breast cancer could be challenging for both patient and physician. Modern bioethical standards require that patients be involved in the process of decision making. Accordingly, the ultimate authorization to start or stop a medical treatment must be based on a comprehensive information process. However it poses a dilemma, how to reconcile patient autonomy with the principle of doing no harm. Solving the theoretical dilemma and implementing it to practice is challenging, but the shared-decision-making model appears to be a promising tool in this respect. Regarding CRRM and BRRM the *Manchester guidelines for contralateral risk-reducing mastectomy* provides a great example of how to complete the solely medical assessment of the patient who requests mastectomy for risk-reductive reasons with ethical considerations. The Manchester guidelines encourages the use of simple and comprehensible language, personalized medical information tailored to the given patient, and if it is medically agreeable the insertion of some waiting time. The latter, we argue is not necessarily a disproportionate limitation of patient autonomy. It is clear that the patient information and informed consent of women who request risk-reduction mastectomy raises complex ethical issues. We propose, that most of these issues can be satisfactorily resolved in the framework shared decision making.

References

[1] Beauchamp T.L., Childress, J.F. *Principles of Biomedical Ethics*. New York: Oxford University Press, 2009.

[2] Elwyn, G., Frosch, D., Thomson, R., Joseph-Williams, N., Lloyd, A., Kinnersley, P., … Barry, M. "Shared Decision Making: A Model for Clinical Practice." Journal of General Internal Medicine 27(10) (2012): 1361-1367.

[3] Jacobson, N. "The socially constructed breast: breast implants and the medical construction of need." *American Journal of Public Health* 88(8) (1998): 1254-1261.

[4] Goldwyn, R.M. "Subcutaneous Mastectomy." *New England Journal of Medicine* 297(20) (1977): 503-505.

[5] Wong, Stephanie M.; Freedman, Rachel A.; Sagara, Yasuaki; Aydogan, Fatih; Barry, William T.; Golshan, Mehra. "Growing Use of Contralateral Prophylactic Mastectomy Despite no Improvement in Long-term Survival for Invasive Breast Cancer." *Annals of Surgery* 265(3) (2016): 581-589.

[6] Evans, D. G., Wisely, J., Clancy, T., Lalloo, F., Wilson, M., Johnson, R., … Howell, A. "Longer term effects of the Angelina Jolie effect: increased risk-reducing mastectomy rates in BRCA carriers and other high-risk women." Breast Cancer Research 17 (2015).

[7] Desai, S., & Jena, A. B. "Do celebrity endorsements matter? Observational study of BRCA gene testing and mastectomy rates after Angelina Jolie's New York Times editorial." *The BMJ* 355(i6357) (2016). http://doi.org/10.1136/bmj.i6357. Accessed: 2017. 07.01.

[8] Evans, D. G. R., Barwell, J., Eccles, D. M., Collins, A., Izatt, L., Jacobs, C., … Murray, A. "The Angelina Jolie effect: how high celebrity profile can have a major impact on provision of cancer related services." *Breast Cancer Research* 16(5) (2014): 442.

[9] Kovacs, J. Fundamentals of Modern Medical Ethics. An Introduction to Bioethics. Budapest: Medicina, 2006.

[10] Chewning, B., Bylund, C., Shah, B., Arora, N. K., Gueguen, J. A., & Makoul, G., "Patient preferences for shared decisions: A systematic review." Patient Education and Counseling 86(1) (2012): 9-18.

[11] Levinson, W., Kao, A., Kuby, A., & Thisted, R. A. "Not All Patients Want to Participate in Decision Making: A National Study of Public Preferences." Journal of General Internal Medicine 20(6) (2005): 531–535.

[12] Strull WM, Lo B, Charles G. "Do Patients Want to Participate in Medical Decision Making?" JAMA 252(21) (1984): 2990-2994.

[13] Livaudais, J. C., Franco, R., Fei, K., & Bickell, N. A. "Breast Cancer Treatment Decision-Making: Are We Asking Too Much of Patients?" Journal of General Internal Medicine 28(5) (2012): 630–636.

[14] Nicolai, J., Buchholz, A., Seefried, N., Reuter, K., Härter, M., Eich, W., Bieber, C. "When do cancer patients regret their treatment decision? A path analysis of the influence of clinicians' communication styles and the match of decision-making styles on decision regret" Patient Education and Counseling 99(5) (2016): 739-746.

[15] Martinez, K. A., Kurian, A. W., Hawley, S. T., & Jagsi, R. "How can we best respect patient autonomy in breast cancer treatment decisions?" *Breast Cancer Management* 4(1) (2015): 53–64.

[16] Bellavance, E. C., & Kesmodel, S. B. "Decision-Making in the Surgical Treatment of Breast Cancer: Factors Influencing Women's Choices for Mastectomy and Breast Conserving Surgery." Frontiers in Oncology 6(74) (2016).

[17] Lam, WWT., Chan, M., Or, A., Kwong, A., Suen, D., Fielding, R. „Reducing Treatment Decision Conflict Difficulties in Breast Cancer Surgery: A Randomized Controlled Trial." Journal of Clinical Oncology 31(23) (2013): 2879-2885.

[18] Waljee, JF., Rogers, MAM., Alderman, AK. "Decision Aids and Breast Cancer: Do They Influence Choice for Surgery and Knowledge of Treatment Options?" Journal of Clinical Oncology 25(9) (2007): 1067-1073.

[19] Angelos, P., Bedrosian, I., Euhus, D. M., Herrmann, V. M., Katz, S. J., & Pusic, A. "Prophylactic Mastectomy: Challenging Considerations for the Surgeon" Annals of Surgical Oncology 22(10) (2015): 3208–3212.

[20] Liu, Y., Pérez, M., Aft, R. L., Massman, K., Robinson, E., Myles, S., … Jeffe, D. B. "Accuracy of Perceived Risk of Recurrence among Patients with Early-stage Breast Cancer." Cancer Epidemiology, Biomarkers & Prevention 19(3) (2010): 675–680.

[21] Baptiste, D. F., MacGeorge, E. L., Venetis, M. K., Mouton, A., Friley, L. B., Pastor, R., … Bowling, M. W. "Motivations for contralateral prophylactic mastectomy as a function of socioeconomic status." BMC Women's Health 17(1) (2017).

[22] Yang, YT., Pike, ER., Rose, CM., Botnick, LE. "The rise in bilateral mastectomies: Evidence, ethics, and physician's role." *The Breast* 29 (2016): 160-162.

[23] O'Neill, O., "Some limits of informed consent." *Journal of Medical Ethics* 29(1) (2003): 4–7.

[24] Savulescu, J. "Autonomy, the good life and controversial choices." 17-37. In *The Blackwell Guide to Medical Ethics*, ed. Rhodes, R., Francis, L., Silvers, A. Oxford: Blackwell, 2007.

[25] Jayaraman, S., Mann, M. "Male and female sterilization." *Obstetrics, gynaecology and reproductive medicine* 22(4) (2012).

[26] FIGO. *Ethical Issues in Obstetrics and Gynecology*. London: Figo House, 2012.

[27] Pereira, J. "Legalizing euthanasia or assisted suicide: the illusion of safeguards and controls." Current Oncology 18(2) (2011): 38–45.

[28] Basu, N. N., Ross, G. L., Evans, D. G., & Barr, L. "The Manchester guidelines for contralateral risk-reducing mastectomy." *World Journal of Surgical Oncology* 7(13) (2015): 237.

[29] Borgen PI1, Hill AD, Tran KN, Van Zee KJ, Massie MJ, Payne D, Biggs CG. "Patient regrets after bilateral prophylactic mastectomy." *Ann Surg Oncol* 5(7) (1998): 603-606.

[30] Rosenberg SM., Partridge AH. "Contralateral Prophylactic Mastectomy An Opportunity for Shared Decision Making." *JAMA Surg.* 149(6) (2014): 589-590.

Haematological Patients' Perception of Their Quality of Life

Alciona Sasu, Mircea Onel, Cristina Ghib-Para,
Ligia Piroş, Alin Greluş, Coralia Cotoraci

Introduction

Haematological patients are usually complex patients, whether they have benign or malignant pathologies. The patients admitted in a Hematology Clinic are mostly oncohaematological patients, with severely altered health status, with multiple and frequent complications. Cancer diagnosis brings profound changes in the lives of the patient and their family. Cancer is a chronic illness that has a major impact on general activities, including social integration during and after treatment. Moreover, chemotherapy changes the general status of the patient and frequently involves patient admission into the hospital. During chemotherapy, the patient might have adverse reactions at an incidence depending on the drug administered and the patient's clinical background [1]. Chemotherapy toxicity can take several aspects, from asthenia to severe haematological, digestive, immunological, or dermatological complications. Moreover, toxicity can be early or tardive, thus influencing the patient's status immediately after (but also months after) treatment [2]. Life-changing diagnosis, modified life style, acceptance of diagnosis, psychological disturbances, prolonged hospitalisation, privation of family life and social life -- these are only some of the challenges that oncological patients face shortly after diagnosis.

In the last decades, medicine has underlined the importance of quality of life [3], besides the specific indicators which stress patient outcome (e.g. survival rate, disease free survival, etc).

The concept of quality of life is hard to define and should be used with great discernment [4]. According to the World Health Organization, quality of life is given by the individuals' perceptions of their own social situations in the context of the cultural value systems they live in, as well as their own needs, standards, and aspirations [5]. Health related quality of life means physical, mental, social [6] and, more recently, spiritual well-being, as well as the ability of patients to perform their ordinary tasks in their everyday life [7].

Life quality assessment is done using varied fields with multiple indicators.

Different models of quality of life assessment have been designed: some specific to certain pathologies [8], others trying to make a difference between function and health, assessing health in the context of permanent functional limitations [9].

Having all these data, haematologists and oncologists conduct a complex therapy management of their patients, combining the standard ethiologic therapy with the psychological, social, spiritual, or alternative therapies [10].

Material and methods

Patients

We conducted a prospective study over a period of 12 months (2015/2016) within a group of 50 patients admitted in the Department of Haematology of the Clinical County Emergency Hospital of Arad (Romania). They were patients diagnosed with oncohaematological diseases. The patients were divided into 3 lots:

Lot A - patients who underwent oncological treatment (n = 27);

Lot B - patients who underwent oncological and psychiatric treatment (n = 15);

Lot C - patients who underwent oncological and psychological treatment (n = 8).

Hypothesis

In order to achieve the goal, we tested the following hypothesis:

Hypothesis 1 – physical health affects mental health status, social behaviour, and relationships;

Hypothesis 2 – mental health influences social activity;

Hypothesis 3 – oncological, psychiatric, and psychological treatment influences the state of mental health, general health, and quality of life of oncohaematological patients;

Hypothesis 4 – the emergence of a stress factor before the onset of illness affects mental health.

Objective

The main objective: Assessing the quality of life in oncohaematological patients and highlighting the emotional, physical, and social factors impacting on the indices that characterize the quality of life of these patients.

Working documents

1. Informed consent;

2. Anamnestic test;

3. Special psychological tests for quality of life assessment: MOS-SF-36 - Medical Outcome Study-Short Form 36 (principal author John Ware Jr., 1992, having 8 fields: Physical functionality scale, Scale of problems caused by physical disorders, The social functionality scale, Scale of body pain, The mental health scale, Scale of problems caused by emotional states, Vitality scale, The general health scale) and SCL-90-R Symptom Checklist Revised (Derogatis, 1983, with 9 items: Somatization, Obsessive-compulsive disorder, Anxiety, Depression, Interpersonal sensitivity or sensitivity, Hostility, Phobic anxiety or phobias, Psihoticismus, Paranoid ideation).[1]

Results

The study included 50 subjects aged between 42 and 80, with an average age of 65.08 years; 50% were women and 50% were men; 44% from urban areas and the remaining 56% from rural areas; 58% had elementary studies, 8% of subjects were without education, 16% had secondary education, and 18% had higher education. Concerning the income, 32% had a lower income than the mean economy income, 56% had a minimum per economy income, and 12% had an average per economy income.

The majority of 84% lived with their family, 16% lived alone, and no subject was institutionalized.

Of all participants, 32% had a history of psychiatric disease and the remaining 68% had no such history. Moreover, 4% of the subjects had a history of psychological trauma and the remaining 96% had no previous psychological trauma.

Of the study subjects, 6% often drank alcohol, another 16% occasionally drank alcohol, and the vast majority of 78% did not drink alcohol.

The group structure based on behaviour resulted from the question "What is your activity during the day?". The responders answered as following: 7.6% preferred social withdrawal, 52.8% had minimum social activity and 39.6% were socially active.

Concerning the therapy used for the patients, 54% only had oncological treatment, 30% had oncological treatment associated with psychiatric treatment, and 16% had oncological treatment associated with psychological therapy (Fig 1).

---

[1] The authors used authorized sources for all these tests.

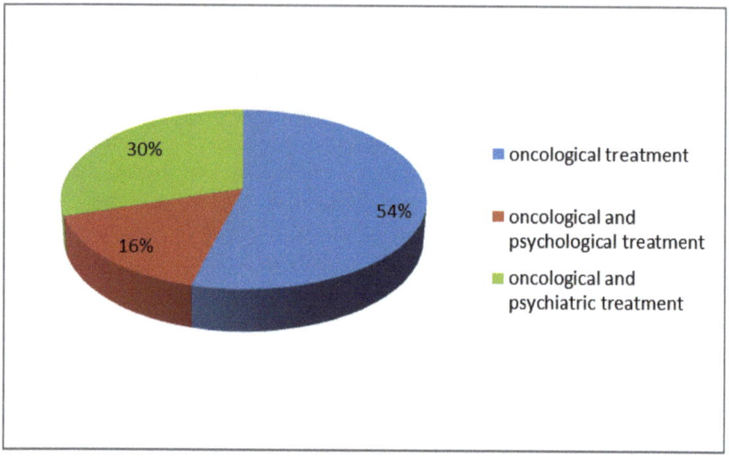

Fig. 1. The group structure according to the treatment

Descriptive analysis of the test and anamnestic results on the 4 plans

- Physical-biological plan

Somatization: reflects the distress which occurs as a result of the perception of bodily dysfunction, the symptoms being centred on the gastrointestinal, respiratory, muscular discomfort (Table 1). The higher the scores, the worse the quality of the person's life.

Table 1. Somatization according to SCL 90

N	50
Mean	26.4800
Std. Deviation	6.97880
Minimum	14.00
Maximum	44.00

- The general health scale

Test SF 36: Score 5-25, the higher the scores, the better the quality of life (Table 2).

Table 2. General Health Scale according to Questions 1, 2 and 11 (5- 25)

N	50
Mean	15.1600
Std. Deviation	2.78010
Minimum	9.00
Maximum	21.00

In order to determine the general health level so that we can divide the participants into two groups, we calculated a general health score based on the answers to questions 1, 2 and 11 of SF. 36.

Subjects with a lower score than the median per group were included in the group of those with a lower health status, and those with a higher than median score were included in the group of those with a better health status.

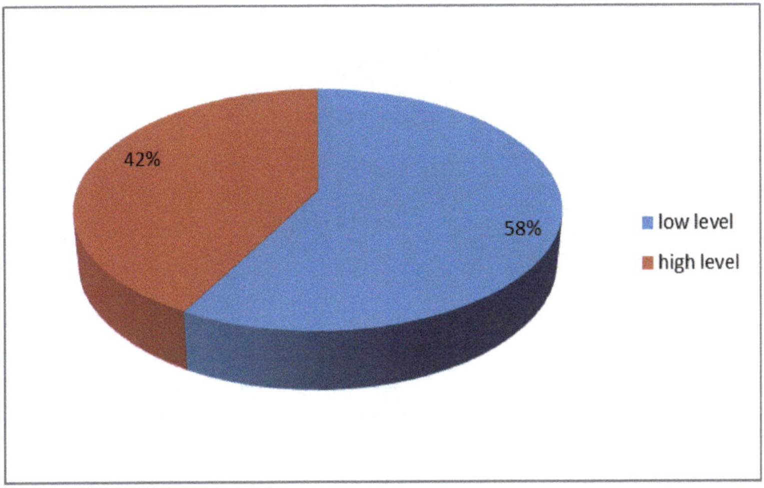

Fig. 2. Group structure according to the general health status

- Physical functionality scale: was described from question 3 SF 36, scores 10-30 (Table 3). The higher the scores, the better the quality of life.

Table 3. Physical functionality scale according to SF 36

N	50
Mean	17.1600
Std. Deviation	5.28903
Minimum	10.00
Maximum	29.00

- The social functionality scale: was described from question no. 6 of the SF 36 test (Fig 3): "In the last 4 weeks, to what extent have your physical health or emotional problems affected your ordinary activities related to family, friends, neighbours, or other groups of people?"

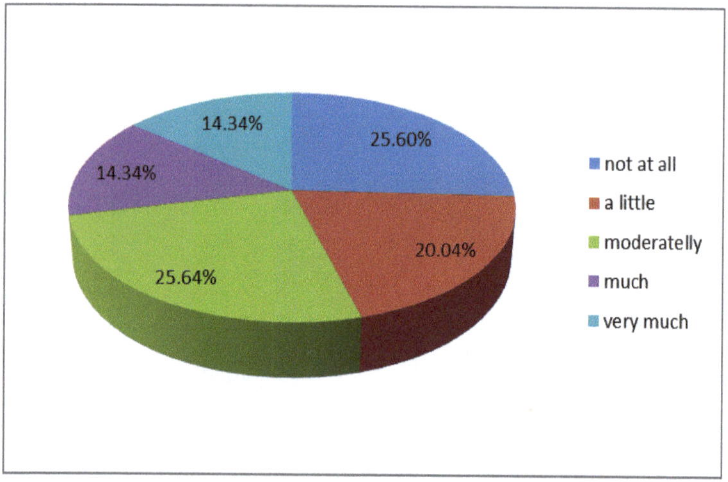

Fig. 3. Group structure according to social functionality scale

- Psycho-emotional plan

The results showed that 60% of the subjects had a happy childhood and 40% had an unhappy childhood.

Concerning the existence of an unfortunate event (stressor) before the onset of the disease, the results showed that 66% of the subjects experienced such an event and 34% have not.

For the existence of obsessive-compulsive disorder, anxiety, depression, sensitivity, interpersonal hostility, phobia, psychoticism, paranoid ideation (Scl 90) the results are presented in Table 4. The higher the scores, the worse the quality of life.

Table 4. Group structure according to psychiatric problems

	Obsessive compulsive	Depression	Hostility	Anxiety	Sensitive	Phobia	Paranoid ideation	Psihoticism
N	15	15	15	15	15	15	15	15
Mean	16.133 33	19.333 3	8.6000	8.6000	13.0000	10.266 7	8.6667	12.333 3
Std. Devia -tion	6.0694 4	8.8855 2	3.0659 4	6.6811 7	5.51621	5.5093 0	3.3735 7	4.1518 8
Min.	10.00	12.00	6.00	10.00	9.00	7.00	6.00	10.00
Max.	26.00	34.00	14.00	31.00	25.00	27.00	15.00	23.00

- The level of mental health

In order to determine the mental health level so that we can divide the participants into two groups, we calculated a mental health score based on the score of the Scl 90 test. Subjects with a lower score than the median per group were included in the group of those with a higher level of health and those with a higher than median score were included in the group of those with a lower health level. The results showed that 50% had a low level of mental health and 50% a had high level of mental health.

Comparative analysis of the testological and anamnestic results

- Impact of overall health on the social plan (Fig. 4)

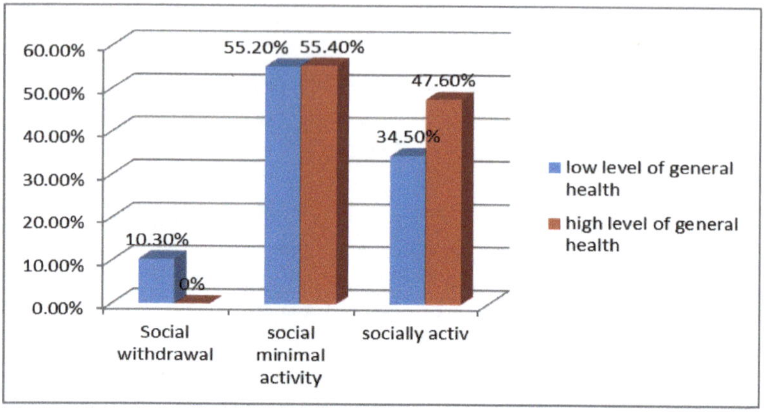

Fig. 4. The influence of physical-general health on behaviour

- The influence of the physical-general health state on the social function: was deduced from the question: "In the last 4 weeks, did you have moments when your physical or emotional health has affected your social activities (e.g. visits to friends, relatives etc.)?" (Fig. 5).

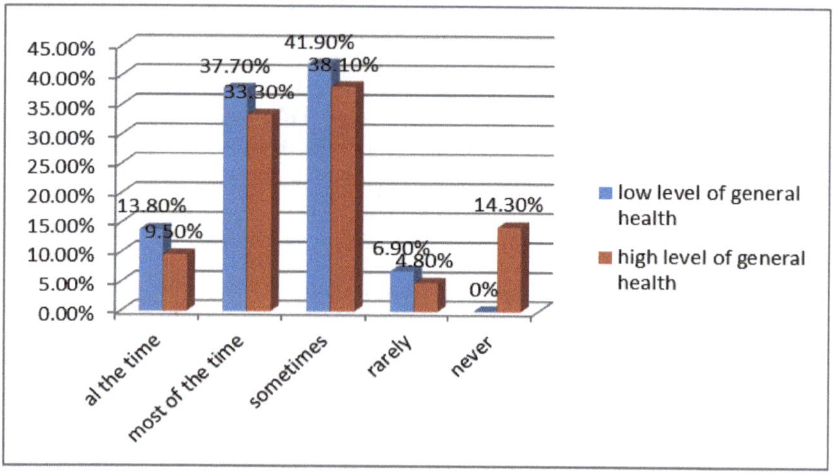

Fig. 5. The influence of physical-general health on social function

- The impact of mental health on the social plan (Fig. 6)

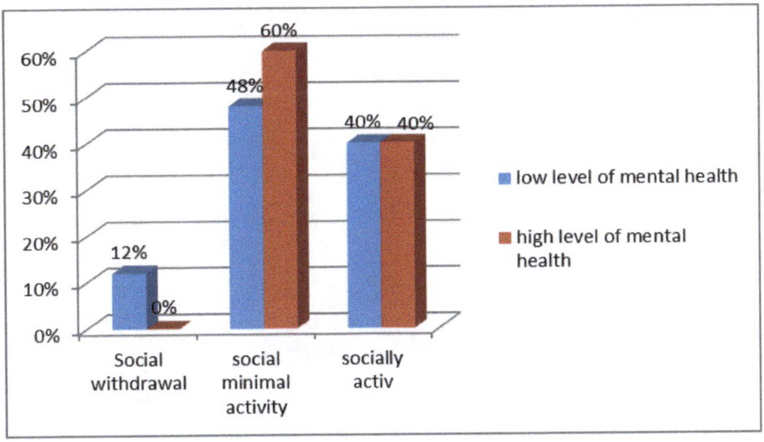

Fig. 6. The influence of mental health on behaviour

- The influence of mental health status on social function: was deduced from the question "In the past 4 weeks, did you have moments when your physical or emotional health has affected your social activities (e.g. visits to friends, relatives, etc.)?" (Fig. 7).

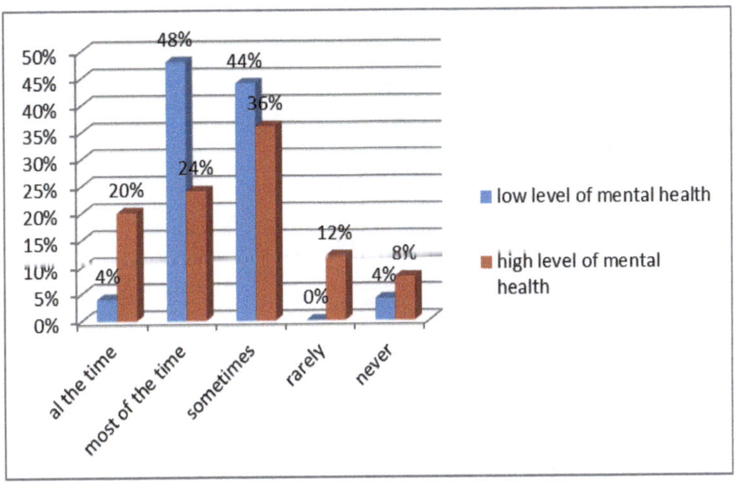

Fig. 7. Influence of mental health status on social function

- The influence of treatment on mental health (Fig. 8)

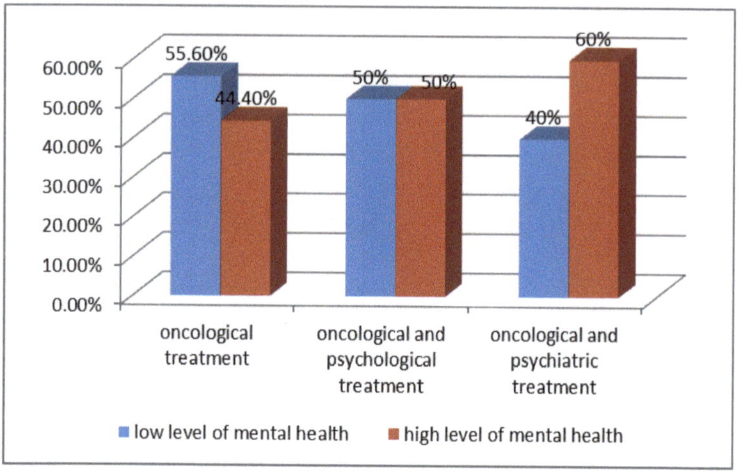

Fig 8. The influence of treatment on mental health

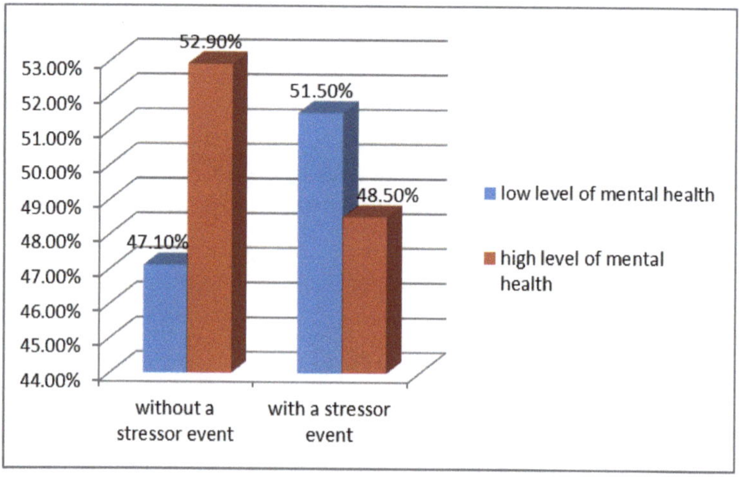

Fig. 9. The influence of the presence of an unfortunate (stressor) event before the onset of illness, on mental health

Discussion

Quality of life is a topic which has become increasingly important in recent years, demonstrating the need to extend clinical research and practice beyond simply assessing the size of symptoms. Furthermore, it has shown that the goal of treatment is not only reducing symptoms but also improving the quality of life [11].

Among the most important factors influencing the quality of life in the subjects of this study were the biological factors in which we can intervene by treatment (depression, anxiety, somatic pathology, oncological pathology). Psychiatric comorbidities were present in 32% of male and female subjects. 4% of the subjects suffered physical trauma in their past, which is a factor that reduces quality of life. Biological factors (somatic and psychiatric) can be successfully intervened upon, thus improving the patients' quality of life.

As stated before, of the study subjects, 54% had only oncological treatment, 30% had oncological treatment with psychiatric treatment, and the remaining 16% had oncological treatment with psychological treatment (Fig 1). Recent studies support the hypothesis that when combined with chemotherapy, psychiatric therapy can improve the effects of chemotherapeutic agents, suggesting that SSRI (selective serotonin reuptake inhibitors) antidepressants have the potential to chemosensitize cancer cells in chemotherapy regimens [12-13].

Psychological factors

The psychological state of cancer patients is correlated with the stages of the disease: there is a decrease in depression and anxiety rates when patients are in the remission phase and they increase in the relapses of the disease. The stigmatization of psychiatric and oncological pathology can lead to a demonstrated decrease in patient compliance, which falls in a vicious circle: each new recurrence accentuating the stigma of the entourage. Humour, optimism, resilience, memory of past positive experiences, positive suggestiveness, internal controlling site (assuming responsibility for monitoring the first symptoms of a new episode, adherence to drug treatment) are factors which can be cultivated through psychological education (16% constantly participating in psychotherapy programs) in order to determine the most favourable development of the underlying disease and to improve the quality of life [14].

Socio-economic factors

Social support can be granted by family, colleagues, friends, but also by institutions such as church or various forms of social protection. Social support plays a protective rôle concerning stressors predisposing to depression and anxiety (Table 4) [15].

Among the study subjects, the majority of 84% live with family and the remaining 16% live alone. Of the participants in the study, 7.6% prefer loneliness, 52.8% have a minimum social activity and the rest of 39.6% are socially active. The subjects have proved to have good social support, the majority living with the family and contributing to the improvement of the quality of life.

Somatization reflects the distress which occurs as a result of the perception of bodily dysfunction, accusations being centred on gastrointestinal, respiratory, and muscular discomfort.

The scores on this group of subjects (Table 1) ranged from 14 to 44, averaging 26.48, concluding that subjects with psychiatric comorbidities have a very low perception of quality of life.

Of the study subjects, concerning the mental health, 50% had low mental health level and the rest high level of mental health. Quality of life is inversely proportional to the score obtained from the Scl 90 testology for subjects suffering from psychological disorders, with approximately half of the subjects achieving high scores following the test.

The test being applied during hospitalization, at the time of crisis of the subject, reveals that this period is a critical moment in which the declared quality of life is close to the minimum.

The scores obtained in the SF 36 physical scale on this group of subjects ranged from 10 to 29 with an average of 17.16, which demonstrates that in general the oncohaematological patient's functionality is influenced, the quality of life being altered due to this aspect.

The scale of social functionality applied at the time of hospitalization has better quality of life scores, only one-third of subjects feel the disease has affected their relationship with those around them.

The data obtained helps conclude that childhood contributes to the mental health of the subject and, implicitly, to the quality of life of the subject.

It was observed that subjects who had a stressor before the onset of the disease (66%) had a lower level of mental health than those who did not have such an event (34%).

Conclusions

The data of this study shows that *physical health* influences the state of mental health, behaviour, and relations on the social plan. For all the mental health dimensions of the Scl 90 test, individuals with low physical health have been found to have scores higher than those with a high level of physical health, which means that people with reduced physical health are psychologically affected more than those with a good physical health. Moreover, subjects with a high level of general physical health are more socially active than those with a low level of general physical health. It has been observed that subjects with a high level of general physical health have better relationships with people surrounding them than those with a low level of general physical health.

Mental health influences behaviour, social function, and relationships with people around. Subjects with a low mental health level are more socially-withdrawn than those with an increased mental health level (Fig. 6, 7). Subjects with an increased mental health level have better relationships with people around them than those with a low mental health level.

Oncological, psychiatric, and psychological *treatment*, influences the state of mental health, general health, and quality of life of oncohaematological patients. Subjects who have undergone psychiatric and psychological treatment together with oncological treatment have a higher mental health level than those who have only had oncological treatment (Fig 8).

The emergence of a *stressor factor* before the onset of the disease, influences mental health. Subjects who had a stressor before the onset of the disease have a lower level of mental health than those who did not have such events before the onset of the disease.

It is clearly shown that physical status and emotional status influence the patient's quality of life. There are also other keys – such as the stressor event before the onset of the disease – that should evaluated more deeply. Moreover, the mental status of the patient resulted from his history can be relevant when appreciating the psychological influence of cancer on the patient's emotional status. The patient's family plays an important role in patient's social quality of life.

This study helps in the complex evaluation of an oncohaematological patient, highlighting the main emotional and social factors which affect the indices which characterize the quality of life of oncohaematological patients in order to develop strategies to prevent and improve it.

References

[1] Deepti Chopra, Harmeet S. Rehan, Vibha Sharma, and Ritu Mishra, "Chemotherapy-induced adverse drug reactions in oncology patients: A prospective observational survey", *Indian J Med Paediatr Oncol.* 2016 Jan-Mar; 37(1): 42–46. doi: 10.4103/0971-5851.177015 PMCID: PMC4795375

[2] Ajitha Sharma, K. Meena Kumari,1 Hasitha Diana Manohar,1 K. L. Bairy,1 and Joseph Thomas, "Pattern of adverse drug reactions due to cancer chemotherapy in a tertiary care hospital in South India", *Perspect Clin Res.* 2015 Apr-Jun; 6(2): 109–115. doi: 10.4103/2229-3485.154014, PMCID: PMC4394577

[3] Pennacchini M, Bertolaso M, Elvira MM, De Marinis MG "A brief history of the Quality of Life: its use in medicine and in philosophy", *Clin Ter.* 2011;162(3):e99-e103

[4] Marcel W.M, "Definitions of Quality of Life: What Has Happened and How to Move On", *Top Spinal Cord Inj Rehabil.* 2014 Summer; 20(3): 167–180. Published online 2014 Aug 11. doi: 10.1310/sci2003-167, PMCID: PMC4257148

[5] WHOQOL Group. Development of the World Health Organization WHOQOL-BREF quality of life assessment. *Psychol Med.* 1998;28:551–558. [PubMed]

[6] http://www.healthmeasures.net/explore-measurement-systems/promis/intro-to-promis

[7] Bottomley, Andrew, "The Cancer Patient and Quality of Life", *The Oncologist*, 7 (2): 120–125. ISSN 1083-7159. PMID 11961195. doi:10.1634/theoncologist.7-2-120. Retrieved 2015-05-03. (April 2002).

[8] Marcel W.M. Post, "Definitions of Quality of Life: What Has Happened and How to Move On", *Top Spinal Cord Inj Rehabil.* 2014 Summer; 20(3): 167–180. Published online 2014 Aug 11. doi: 10.1310/sci2003-167, PMCID: PMC4257148

[9] Krahn GL, Suzuki R, Horner-Johnson W, "Self-rated health in persons with spinal cord injury: relationship of secondary conditions, function and health status", *Qual Life Res.* 2009 Jun; 18(5):575-84

[10] Katarzyna Cieślak, "Professional psychological support and psychotherapy methods for oncology patients. Basic concepts and issues", *Rep Pract Oncol Radiother.* 2013 May; 18(3): 121–126. Published online 2013 May 3. doi: 10.1016/j.rpor.2012.08.002 PMCID: PMC3863260

[11] Krouse, H. J., "Efficacy of video education for patients and caregivers", *ORL-Head &Neck Nursing*, no. 21(1), 2003.

[12] Scuster, C., Fernbach, N., Rix, U., Superti-Furga, G., et al., "Selective serotonin reuptakeinhibitors – a new modality for the treatment of lymphoma/ leukaemia?", *Biochemical Pharmacology*, no. 74(9), 2007

[13] Amit, B. H., Gil-Ad, I., Taler, M., Bar, M., Zolokov, A., Weizman, A., "Proapoptotic and chemosensiting effects of selective serotonin reuptake inhibitors on T cell lymphoma/ leukemia (Jurkat) in vitro", *European Neuropsychipharmacology*, no. 19(10), 2009

[14] A M David, S Nastase, CALITATEA VIEȚII, XXIII, nr. 1, 2012

[15] Dobson, K., *Handbook of cognitive-behavioral therapies* (2nd ed.), New York, The Guilford Press, 2001.

PART IV

END OF LIFE ETHICAL ISSUES

Is Death the Enemy? The Normative Power of Metaphor in Bioethics

Assya Pascalev

"The greatest thing by far is to be the master of metaphor."
Aristotle, Poetics, 322 BC

Introduction

Metaphor consists in "seeing," understanding or speaking of one thing in terms of another,[1] i.e., "Man is a wolf," "Time is money," "All the world's a stage." Traditionally, metaphor has been construed in purely linguistic terms as a powerful rhetoric device intended to convey images and emotions, and to persuade the listener.[2] Outside literary discourse, however, metaphor has been viewed as unnecessary, dispensable and even dangerous because of its ability to "insinuate wrong ideas, move the passions, and thereby mislead the judgment; and therefore, wholly to be avoided" [2, p. 34]. This understanding of metaphor as ornamental and expendable coupled with the belief that metaphor has no place in serious fields such as science and philosophy has resulted in a lasting neglect of the role of metaphor outside literary discourse. As a result, philosophers have paid little attention to metaphor turning a blind eye to the presence of numerous metaphors in philosophical theories. Similarly, ethicists have largely failed to examine the role which metaphor plays in moral reasoning and, with a few exceptions, the metaphors, which operate in medicine and bioethics have received little philosophical scrutiny.[3] Recent studies in linguistics and cognitive science challenge this received attitude by

---

[1] In this most general definition, the word "thing" refers to any object, including physical objects, events and states of affairs.

[2] In *Rhetoric*, Aristotle states that metaphors "set the scene before your eyes" [1] (1410), enable one to "get hold of new ideas" [1] (1410b) and give "style clearness, charm and distinction as nothing else can" [1] (1405a).

[3] For a detailed analysis of the metaphors underlying Western philosophical thought, see [3] (Part III).

showing that metaphors permeate language and play an important role in thought and action: metaphors operate in everyday language, science, philosophy and religion, and they function as conceptual frameworks, which shape the way we understand reality and our place in it [3, 4, 5]. This new conception of metaphor as a powerful instrument for understanding and even structuring reality, our perceptions and attitudes calls for an examination of the nature and functions of metaphor in fields traditionally considered metaphor-free, including mathematics, the natural sciences, medicine, philosophy and bioethics.

The purpose of this chapter is to explore the role of metaphor in biomedical ethics and its effects of the moral and professional conduct of physicians. It advances the view that metaphor has an important normative (action-guiding) function in moral reasoning and action, and that metaphors serve as vehicles of moral knowledge by allowing us to understand the values, concepts and principles associated with a particular metaphor. The focus of this chapter are the metaphors of death and dying employed by physicians in end-of-life care. I argue that physicians conceptualize the phenomena of death and dying through the use of metaphors, which have a normative function and prescriptive power. I identify three dominant metaphors of death and dying employed by physicians: the WAR, the TORTURE and the ART metaphors.[4] I explicate the normative assumptions behind each metaphor and show that the metaphors influence physicians' attitudes towards the dying, shape physicians' moral and professional reasoning, structure their experience and influence medical decisions concerning the kind and quality of care at the end of life. Firstly, I analyse the WAR metaphor, which represents death as the chief enemy of medicine. The metaphor highlights the absolute value of life, encourages aggressive treatment and the use of medical technology regardless of the likely outcome. The metaphor ignores the patient's physical discomfort, and the psychological and emotional cost of the interventions. Next, I discuss the TORTURE metaphor representing dying as a technologically-aided torture. The metaphor emphasizes the patient's perceived quality of life and downplays the role of therapeutic intervention. The metaphor establishes the moral and psychological basis for limiting treatment. Lastly, I discuss the metaphor of DYING AS AN ART. It characterizes dying as the finale of a composition, and end-of-life care as orchestration. I conclude that in clinical settings, moral conflicts at the end of life are understood and resolved by the use and shift of metaphors of death and dying. The metaphors reassign moral responsibility, shape treatment decisions and change physicians' attitudes to the dying.

---

[4] The three dominant metaphors of death and dying are identified and analysed at length in my dissertation *Medical Metaphors of Death and Dying: An Ethical Analysis* [6].

What is a metaphor and why is it ethically significant?

The word "metaphor" comes from the Greek term *metaphora*, which means "transfer" [7, p. 896], [8, p. 1]. Metaphors have a distinct basic structure with two components: one is the thing we are trying to understand or characterize, the other--the thing in terms of which we understand or characterize the first object.[5] In philosophy, there exist two major positions on the nature and role of metaphor. The ***received view***, which dominates most of the history of philosophy until the 20th century, holds that metaphor is merely a figure of speech with purely decorative function. It lacks cognitive value and ought to be avoided in rational discourse. In the 20th century, under the influence of cognitive science and linguistics, an alternative view emerges. This view called ***constructivism*** construes metaphor as essential to language and thought, and argues that it has important cognitive functions. The proponents of constructivism argue that metaphors are devices, which structure our experience, and help us to understand the world and our place in it. The constructivist view of metaphor could be traced back to the philosophical stance of I. Kant and F. Nietzhsce. However, it was articulated most clearly in the works of 20th century rhetorician I. A. Richards and philosopher Max Black and further developed by linguist John Lakoff and philosopher Mark Johnson. Lackoff and Johnson's seminal books *Metaphors We Live By* [4] and *Philosophy in the Flesh* [3] analyze a huge number of metaphors, including the metaphors used in philosophical discourse. Drawing on research in cognitive science, Lackoff and Johnson (hereafter L&J) show that metaphors function as conceptual frameworks, which can influence human actions, change cultural values, and shape social institutions.

It is beyond the scope of this chapter to undertake a detailed analysis of L&J's view. For our purposes, it suffices to outline their central arguments for the normative significance of metaphor. L&J begin by observing that

---

[5] The prevalent two-component view of metaphor has been criticized by some as too restrictive and inadequate. Opponents such as J. Soskice argue that the two-component structure is present only in metaphoric statements of the form "X is Y" but not in metaphors with different syntactic forms. Soskice cites counterexamples such as "The pages were covered with a writing script," "The children flew to the window," "He went about in a nimbus of despair" and phrases like "darker purposes," "rosy-fingered dawn," "bitter disappointments," and "sweet voices." She maintains that these are metaphoric expressions which lack two distinct components and that "to eliminate instances like these because they lack two explicit subjects would be to eliminate most of what we take to be metaphor" [8, p. 20]. Soskice notes that even those who subscribe to the two-components view, like S. Ullmann and M. Black, include expressions like "bitter disappointments" and "sweet voices" in their lists of metaphors [8, p. 20].

metaphors are ubiquitous in all spheres of activity, including science and philosophy. We speak of time in terms of money ("You are ***wasting*** my time," [4, pp. 7-8]; we view love as a journey ("The marriage is ***on the rocks***," "We are ***stuck***" [4, pp. 64-68] and we discuss theories and arguments as if they are buildings, e.g., "We need ***to construct*** a strong argument," and if we fail, then "the argument collapses" [4, (p. 46]. Philosophical discourse also abounds with metaphors: recall Plato's metaphors of the Good as the sun, knowledge as light, and ignorance as darkness in the famous Allegory of the Cave. Descartes also uses a metaphorical conception of ***knowledge as seeing*** and Kant speaks of the moral community as ***the Kingdom of Ends***.[6] L&J note that our most basic categories such as action, causation, time, and self are metaphoric in nature as is our ordinary conceptual system. The latter structures "what we perceive, how we get around in the world and how we relate to other people" [4, p. 3]. As part of our conceptual scheme, metaphors "structure how we perceive, how we think and what we do" [4, p. 4].

If L&J are correct and our conceptual system is indeed largely metaphoric in nature, then it follows that metaphor has also normative significance: "In all aspects of life… we define our reality in terms of metaphors and then proceed to act on the basis of the metaphors" [4, p. 158]. Since our intentional actions are guided partly by the features of the world, and partly by our goals, commitments and plans, then the metaphors, which underlie our understanding of the world and our perceptions, inferences, and plans also shape our actions and goals, and influence our patterns of deliberation. L&J identify the specific cognitive mechanism, which brings about the normative power of metaphor. It results from the interaction between the two central components of metaphor. L&J note that metaphor necessarily has two components: (1) a target domain, which we try to understand, and (2) a source domain, from which we borrow images and experiences in terms of which to understand the target domain. Thus, the "time is money" metaphor emerges when the target domain (time) is connected in a systematic way to the source domain (money, limited resource, valuable commodity). When we use a metaphor, we map one concept onto another forming a complex network of relations. The source domain consists of concepts, phenomena and experiences that are familiar and form a somewhat coherent whole. We use our knowledge about the source domain to understand the phenomena in the less familiar target domain. Relying on our knowledge of the source domain, we draw inferences about the target domain [4]. Consequently, we come to think of the target domain in terms of the objects, properties and relations of the source domain. The metaphoric expressions connect the characteristics of

---

[6] For a detailed analysis of the metaphors underlying Western philosophical thought, see [3] (Part III).

the target and source domains in non-arbitrarily way. These non-arbitrary connections L&J call "entailments." Thus "money" entails "a limited resource" which in turn entails "a valuable commodity". When we transfer the logic of the source domain onto the target domain, we end up with linguistic expressions such as "wasting time," "saving time" and "budgeting time." These expressions are structured by and make sense only in light of the chosen source domain. The latter act as a constraint on how we think about time, what we do with it ("borrow", "save", waste") and how we come to experience it. L&J call this process "metaphoric mapping and structuring." The source domain serves as a frame of reference for understanding the target domain, and the pattern of ordering of metaphoric connections follows the structure and logic of the source domain. In such a way, a metaphor picks out features within the target domain which features would be deemed relevant on the basis of the elements or relations relevant in the source domain. The logic of the source domain highlights some aspects of the target domain, and hides others. When we think of time as money, we ***do not think*** of time as something else, i.e. TIME-IS-ART. The network of metaphoric entailments either fits our experiences, or fails to do so. If there is a fit, then the metaphor is deemed "apt," and the experience is organized according to the metaphoric entailments.

Whether or not a metaphor fits our experience is a complex question because one's experience is shaped by a multitude of factors including one's neurological and bodily constitution, culture and social institutions. Our culture and institutions in turn provide the context, in which some metaphors and ways of thinking would fit while other ways of thinking about those domains would be ruled out. The relation between metaphor, social reality and how we think about and understand this reality points to the elements of control and power inherent in the more complex metaphors, making metaphor a normative matter and, thus, a matter for ethical and political analysis. A metaphor can be used to exert power, to manipulate thought, and to misguide action because "the people who get to impose their metaphors on the culture get to define what we consider to be true—absolutely and objectively true" [4, p. 160].

The constructivist understanding of metaphor is particularly relevant to ethical inquiry, moral reasoning, and bioethics. It suggests that metaphors have normative import because of their capacity to shape moral concepts and to constraint cognition and moral reasoning. By the same token, metaphors become vehicles of moral knowledge and moral change because they allow us to deconstruct the values, concepts and principles associated with the use of a particular metaphor. Last but not least, the conceptual view of metaphor shows that although metaphor is constitutive of thought and understanding, any particular metaphor is in principle open to criticism and change because it is a human construct, which is contingent and need not make sense outside our Western conventions and institutional practices.

The analysis so far shows that understanding the role the metaphor plays in morality and evaluating the specific metaphors we employ becomes a legitimate task of ethical inquiry. Yet, there are very few attempts to identify and discuss medical metaphors from a bioethics perspective.[7] In what follows, I draw on the constructivist theory of metaphor to identify the dominant medical metaphors of death and dying and use them as a vehicle to understand the values, interests and commitments of medical practitioners.

I approach the medical metaphors of death and dying from the perspective of the medical workers in intensive care because, in contemporary American society, about 50% of Americans die in hospitals under intensive care [12]. This makes the ICU the place where both patients and physicians tend to encounter death. My analysis utilizes a variety of sources including an in-depth qualitative study of ICU by medical sociologist Robert Zussman [13] and my own observations at two leading medical institutions. My analysis of the philosophical and cultural roots of the metaphors relies on the works of French historian Philip Aries [13, 14], anthropologist I. Illich [16] and philosopher R. Momeyer [17]. I also draw on the studies of medical sociologists Claudine Herzlich and Janine Pierret [18].

The empirical material reveals three different sets of images of death and attitudes to dying employed by the medical professionals in the ICU. They are organized around three distinct metaphors: the WAR metaphor, the TORTURE metaphor and the ART metaphor. I isolate the target and source domain of each metaphor, I analyse its entailments and their normative consequences, and assess the metaphors from a patient-cantered perspective.

---

[7] In an article titled *Metaphors and Models of Doctor-Patient Relationships: Their Implications for Autonomy*, bioethicists James Childress and Mark Siegler identify five dominant metaphors (or models) used to characterize the physician-patient interaction in bioethical literature. The authors indicate that the metaphors play a descriptive as well as prescriptive role and need to be assessed critically [9]. In his book *The Foundations of Bioethics*, Tristram Engelhardt discusses the metaphor of the captain of the ship as a description of the traditional role of the physician [10]. More recently, Swedish scholar Anders Nordgren has argued for the significance of metaphors in bioethics [11]. Except for these sporadic efforts, bioethicists and philosophers have not explored systematically the ways in which metaphors influence medical practice and their effects of on medical decision-making and the moral deliberation of physicians are still to be explored.

The WAR Metaphor of Dying[8]

The Structure and Meaning of the WAR Metaphor

The WAR metaphor represents the chronologically first medical response to the threat of death in the ICU. It conceptualizes critical care as warfare and death as the enemy of the physician. The gist of the metaphor is captured in the statement of an ICU physician: "we believe in the cure, and we look at death as the enemy and we are gladiators in the battle against disease..." [13, p. 107]. The source domain is war, the target domain is death in the ICU. The statement reveals a particular perception of death as the biggest foe of medicine. Treatment decisions and medical interventions are described in militaristic terms, e.g., physicians administer "batteries of tests," and prescribe "armies of antibiotics." The WAR metaphor exemplifies physicians' active, even aggressive attitude to the challenges of life-threatening illness and death, which gives rise to a militaristic rhetoric. This mode of conceptualizing death and dying has far-reaching consequences for the patients. Zussman writes: "First, virtually all patients are treated actively in the intensive care unit at least at the time of their admission" [13, (p. 107]. All resources are mobilized and aggressive methods of treatment are commonly applied. As a result of this professional response, the patient is likely to receive maximum attention from a qualified staff; she can benefit from the most advanced medical technologies and has better chances for survival. Even if the patient has a bad prognosis, her life can at least be prolonged. The WAR metaphor reveals a peculiar, yet characteristic mixture of associations among different notions and images. The meaning of cure, illness and death are intertwined and disease talk is conflated with death talk. Such a framing of death and dying is deeply rooted in Western culture.[9] According to French historian Aries, it is a distinct feature of modern medicine that death becomes a part of "the disease of which it is one of two

---

[8] I offer a detailed reconstruction of the WAR and TORTURE metaphors in [19, pp. 219-236].

[9] More particularly, in the U.S.A., death and the medical profession have become permanently linked together in medical practice and the public's perception since the late 19th and early 20th centuries. Such a link becomes possible when the institution of hospital emerges as the proper locus of both medical treatment and death. It offers to physicians optimal conditions for applying complex and technologically advanced procedures which are impossible to accomplish outside the institution. Thus, the clinical environment significantly enhanced physicians' ability to combat illness, pain and suffering, and ultimately death. Therefore, in the hospital, the struggle with death becomes the natural continuation of the therapeutic efforts of physicians. As a result, treatment and dying were institutionally conjoined. From that moment on, death became part of the province of medicine and today medicalized death is practically the only mode of dying faced by individuals in the U.S.A.

possible outcomes, the other being recovery" [15, p. 358]. The death of the patient comes to be equated with a medical failure. This image of death as the enemy is evoked not only by the medical staff, it is adopted also by the patients. They use the WAR metaphor, when trying to communicate their experiences in the unit. Thus, a patient who underwent intubation explains: "... all these doctors are taking care of you, you know. They are *fighting* for your life. They don't know you, but they're there *fighting* for your life" [13, p. 94]. The patient's account illustrates the pervasive image of death as the enemy and the power of medical discourse to penetrate the patterns of thinking of lay people. The institutional mission of the ICU determines the framework within which the image of death as an enemy is construed and reinforced: as in war, the ICU interventions take place after the outbreak of a crisis--a life-threatening or acute illness. For the ICU staff, death comes to be perceived as the unwanted result of an organ-system failure, against which the staff consolidates its efforts. Thus, the image of death as an enemy emerges out of the very mission of intensive care. As in wartime, the ICU staff is "armed" with complex equipment which is impossible to access outside the unit. Here, the most advanced and sophisticated medical technologies are constantly applied. Vasopressors, ventilators, apparata for haemodialysis are only a small portion of the intimidating armaments employed in the combat with death. But, perhaps the most eloquent evidence for the militaristic profile of the ICU is its genesis. The emergence of ICUs is the result of earlier efforts of American military officers to secure medical care for "the more seriously ill but salvageable soldiers."[10] Zussman characterizes the ICU is the subunit in a hospital, where the active orientation of American medicine reaches its high point. The following statement of an ICU resident illustrates this perception: "You send a patient to the [ICU] to escalate the battle, so this is the next logical step in that battle" [13, p. 107].

The image of death as an enemy can be seen as a manifestation of the profound anxiety that we, human beings, experience, when faced with the fact of our mortality. This negative perception is part of the more general value orientation typical of the 20th century Western society manifested in what Illich calls "the total war to all death-dealing agencies" [16, p. 202]. While in the past, people in the Western world felt at one with nature and saw death as part of it, from the 17th century on, a rapid rise of science and technology took place bringing with it the rise of rationalism and its "faith in progress and its triumph over nature" [15, p. 608]. Death, which was associated with nature, began to be viewed as savage, fearful and evil [14, p. 56]. It is in this modern, death-denying sentiment that the socio-cultural roots of the WAR metaphor can be

---

[10] This view on the genesis of Intensive Care Unit is suggested in Thompson [20, p. 225].

found.

The new understanding of death as natural led to the accumulation of medical knowledge and professional skills by physicians.[11] Consequently, the abilities of doctors to control the outcomes of specific diseases, and to provide cure and prolong the lives of their patients, increased. Physicians became an embodiment of the power of science and were admired because of their perceived ability to push away the patient's suffering and death. This view gave rise to the myth that physicians also "had power over death" [16, p. 195]. As a result, death becomes medicalized and it remains medicalized today. From the standpoint of the physician of today, death is perceived as "a kind of medical disaster" [13, p. 37] - an understanding that gives rise to an extreme negative attitude towards death in medical settings.

The Normative Functions of the WAR Metaphor

This metaphor expresses a negative attitude to death and sanctions the active, even aggressive stance towards dying in the ICU. The active appeal of the WAR metaphor corresponds to the humane mission of the medical profession, which is fundamentally oriented against suffering and pain. Respectively, the challenge for the physician is to balance two distinct professional attributes--compassion and compulsiveness. They correspond to the physician's stable obligations to search for diagnosis, to intervene in the name of treatment, and to alleviate suffering. The ICU environment enhances physicians' actual and perceived ability to combat illness, pain and suffering and creates the hope that it can ultimately also combat death. Consequently, the WAR metaphor is evoked mainly when physicians believe that the patient's life can be saved. The metaphor shapes an expectation that the outcome of the physician's efforts will be positive. By implication, the WAR metaphor connotes that the patient's death is a medical failure. This particular representation of death expresses the self-confidence of health professionals in their ability to accomplish the humanitarian mission of medicine to cure, to protect the patient's life, and to save it, when endangered. Therefore, the WAR metaphor can be interpreted as a manifestation of the self-awareness of the medical profession of its own power. The metaphor embodies the professional ethos of Western medicine bequeathed through the Hippocratic Oath and reinforced by the Judeo-Christian tradition.

---

11 At that time, physicians had at their disposal not only the achievements of various scientific disciplines, but were also able to accumulate specific medical knowledge by performing dissections of human bodies that had been sacred once, but were so no longer.

Critical Evaluation

The attitude of death denial that gives rise to the WAR metaphor has been challenged by many a contemporary scholars. Illich and Momeyer argue that the militaristic rhetoric and the faith in technology create a clinical illusion "that our struggle will result in some sort of abiding victory" over death. This unexamined belief rests on perceived or desired, but not actual professional abilities of physicians. [12] Such views are conducive to excessive medical aggressiveness and encourage "an obsessive kind of medicine conceived as a constant struggle against death" [17, p. 158]. The underlying concerns are that the medical techniques for conquering death are distressing for the patients, too expensive for society, and there exists no strong evidence that they reduce mortality or promote patient's quality of life. [13] A major criticism of the WAR metaphor is that it prompts contemporary medicine to go beyond what patients and society may see as acceptable and reasonable treatment.

The TORTURE Metaphor of Dying

The Structure and Meaning of the TORTURE Metaphor

The TORTURE metaphor expresses the physicians' experiences in treating patients whose condition is seen as hopeless. The meaning of the metaphor is captured in a joke popular in two ICUs studied by Zussman:

> Missionaries are in a tribal land and are captured by the natives. They are brought before the chief and the chief asks one of the missionaries, 'What would you prefer, cheechee or death?' and the first missionary says cheechee, at which point the entire tribe descends on him, ties him to a pole. Each member of the tribe [beats] him. And then the rope [is] tied around his hands, and he is dragged about a mile losing bits and pieces of himself. And [he's] thrown over into a ravine. The other missionary looks horrified at what is going on. And the chief comes back to him and asks him, 'Well, what do you prefer, death or cheechee?' The missionary...says... 'I would prefer death.' The chieftain says, 'Yes, but first a little cheechee' [13, p. 111]

---

[12] Among those contemporary scholars who raise serious doubts regarding the legitimacy of the faith and expectations invested in medical profession is Illich [16]. A more recent critique of the cultural authority of medicine is offered by Herzlich and Pierret [18, Ch. 11].

[13] These criticisms, directed in particular to Intensive Care Unit, are offered in Relman [21, pp. 1252-1253].

"Cheechee" is used interchangeably with the word torture to signify death which is certain, but which must be preceded by extensive treatment. The following explanation of an ICU resident articulates the meaning of the metaphor: "[T]orture is a word used to say that we are putting someone through pain without reason, without reasonable hope that we can reverse the pain he has" [13, pp. 111-112]. Like the WAR metaphor, the TORTURE metaphor is constructed relative to the expected therapeutic results. However, contrary to the WAR metaphor, the image of torture emerges when the foreseen outcome is negative. Zussman observes that, typically, "cheechee" is evoked either (i) when death is certain, or (ii) in cases of incurable patients, or (iii) when "some treatments may simply extend a life that is already wrecked by suffering" [13, p. 110]. Hence, the TORTURE metaphor reflects professional impotence of physicians. It highlights considerations about the patent's perceived quality of life and downplays the role of therapeutic intervention. The bioethics principles behind this metaphor is the principle of non-maleficence (*primum non nocere*: above all, do no harm). The metaphor establishes the basis for limiting treatment and abstaining from the use of medical interventions to prolonging life. The experience of impotence and the image of dying as torture have a profound effect on the self-image of health professionals. In light of the TORTURE metaphor, the technical means of medical profession, which lend to it its sense of efficiency and ground its self-confidence, undergo an intriguing transformation in the imagination of medical workers. The same technological means which physicians saw as instruments of empowerment in the WAR metaphor (tubes, IVs, ventilators) come to be perceived as inquisition devices under the TORTURE metaphor. On the face of it, the metaphor expresses a concern for the patient's suffering. Yet, it is invoked at the end of life when most ICU patients are unconscious, comatose, or in vegetative state. Such patients do not experience suffering. The suffering captured by the metaphor is the suffering, anguish and frustration of the medical staff facing the limit of medicine. This metaphor is at least as much self-reflective as it is about the patient's well-being.

The Normative Functions of the TORTURE Metaphor

The TORTURE metaphor expresses the moral awareness of physicians that a given treatment is futile, and that it inflicts pain. Because of that, the TORTURE metaphor seems to reflect medical considerations that are ultimately patient-cantered. However, the analysis reveals that this is not the case. Zussman notes that since pain is very difficult to measure, neither the doctors, nor the nurses he interviewed "are at all sure how much pain their patients are suffering" [13, p. 112]. Hence, the perception of dying as a torture is not born out of pure compassion. Rather, in the context of intensive care, torture becomes, "a self-reflective concept...Torture refers not just to the suffering of patients, but also to the suffering of doctors and nurses" [13, p. 113].

The TORTURE metaphor reflects the self-reproach of medical workers when they face the limits of medicine marked by the presence of imminent death. At first glance, this frustration seems to stem from a false dilemma that physicians set for themselves. For to cure a patient at the expense of great suffering, or not to cure her does not seem a genuine dilemma, provided the patient is going to die anyway. But what may seem as an exaggerated frustration is, in fact, an expression of a two-fold crisis of the occupational identity of physicians. As Zussman notes "[m]edicine is caught between two sometimes conflicting values: on the one hand, an ethic of intervention and treatment; on the other hand, an injunction to do no harm" [13, (p. 110]. Thus, the image of torture is evoked, when the harmonious coexistence of compulsiveness and compassion is disturbed. It brings in focus two types of problems, which confront the medical profession today. First, the metaphor expresses a tension within the medical profession between the two fundamental imperatives of medicine: to cure (compulsiveness) and to relieve suffering (compassion). The minimal expectation that medical professionals should meet is that "[i]f they cannot always cure, they can at least minimize pain" [13, p. 110]. From this perspective, the TORTURE metaphor expresses the self-reproach of physicians that by subjecting their patients to futile treatments, they go "beyond the point where you should artificially prolong their life... On the other hand, a willingness to withhold treatment in the name of compassion requires a renunciation of the compulsive search for treatment," observes Zussman [13, p. 110]. Consequently, the metaphor has explicit normative connotations. It prohibits medical workers from inflicting what is perceived as uncompensated, and therefore, unjustified suffering. It can be concluded that the TORTURE metaphor communicates a particular moral stance intrinsic to the nature of medicine as a profession.

The other facet of the occupational crisis of which the TORTURE metaphor of dying becomes a symptom, is determined by conditions external to the medical profession. They have to do with the existing social constraints on the professional autonomy and the cultural authority of physicians. By engaging themselves in procedures that artificially prolong the physiological existence of terminally ill patients "beyond any reasonable prognosis," medical workers develop concerns that they are involved in useless activities and their efforts are not properly directed. Yet, they cannot stop a futile treatment only on the basis of their own judgment, because institutional health care is part of the complex network of society. There, the actions of physicians have to comply with the public demand "to get everything done" and with the scrutiny of legislators, patients' families and hospital administrators. Ultimately, the metaphor of dying as TORTURE manifests physicians' awareness that they are not the sole masters in their province. The TORTURE metaphor goes well beyond consideration about patients' interests; it signals the awareness of doctors that they cannot autonomously 'set their own agenda." The TORTURE metaphor indicates a reversal in both the professional and societal perception of death. With the help of medical technology, dying becomes a continual process of discrete medical victories and losses through which the patient as a person vanishes. The TORTURE metaphor can be interpreted as physicians' own reproach to these social and cultural tendencies that strip human death from its personal, spiritual and moral significance.

Critical Evaluation

The TORTURE metaphor embodies an image of death which is incompatible with the WAR metaphor. It tempers physician's activism, brings forth considerations about the patient's perceived quality of life, and engenders willingness on the part of physicians to limit treatment and to abstain from the use of extraordinary means for prolonging the patient's life. The metaphor emphasizes the moral and technical limits to the power of medicine and urges a reconsideration of its professional tasks. The metaphor displaces physicians' commitment to compulsive care with considerations of futility and compassion; aggressive treatment yields to palliative care and physicians' discretion yields to the public demands for patient's autonomy. The image of dying as torture expresses physicians' awareness of the growing societal mistrust of the medical profession and represents the response of contemporary health-care providers to the public control over the practice of medicine in the USA. Last but not least, the TORTURE metaphor urges physicians to look for alternatives to aggressive care at the end of life. It allows them to acknowledge the need for limiting treatment and offers a conceptual framework for justifying the withholding or withdrawal of medical treatment as a legitimate aspect of end-of-life care.

The ART Metaphor of Dying

The Structure and Meaning of the ART Metaphor

The ART metaphor is an attempt to resolve the tensions by re-conceptualizing dying, death, and the proper role of medicine. The metaphor characterizes the process of dying as the finale of a composition and end-of-life care as the orchestration of the finale. It involves multiple performers (the physician, the health care team and the patient's family) and comprises various conflicting elements such as the responsibility to maintain life, the relief of suffering (beneficence), the imperative to respecting the patient's wishes (autonomy) and not to harm (non-maleficence). The physician is viewed as the conductor of the musical piece. She strives to coordinate the tasks of those involved and to harmonize the conflicting elements so as to achieve harmony (the unity of the piece) and to attain the desired end (the finale, an "optimal," "good" or "dignified" death).

The ART metaphor of dying and death is the most recent way of conceptualizing the process of dying and the nature of end-of-life care in the medical and bioethics literature. The earliest instances of the metaphor can be found in the personal narratives of physicians regarding their experiences in end of life care in the 1990's [22, pp. 9, 25, 29, 59]. However, the most explicit articulation of the ART metaphor can be found in Stuart Youngner's article "Orchestrating a Dignified Death in the Intensive Care Unit" [23]. He evokes the ART metaphor in the context of a discussion of the moral and professional responsibilities of critical care physicians in the contemporary clinical environment of the ICU. This environment combines the availability of advanced medical technology allowing physicians to control death and "to postpone it... in almost every instance" [23, p. 1617] with the awareness of growing health care costs, scarce medical resources and a strong emphasis on the patient's autonomy. According to Youngner, "in these circumstances, the critical care physician cannot avoid the responsibility of orchestrating death by balancing factors such as the patient's autonomy and best interests, medical uncertainty and futility, and notions of 'passive' (permissible) and 'active' (forbidden) euthanasia" [23, p. 1617]. The central role in the process of orchestrating death is relegated to the physician, who is the creator of the artwork: "The critical care physician can be likened to a conductor/composer, struggling to balance and blend various clinical, legal, moral, psychological, spiritual, and, increasingly, economic influences to permit patients an optimal death in less than optimal circumstances" [23, p. 1617]. This emphasis on the physician as the composer of the artwork, as opposed to the patient whose life and death are at stake, begs the question whether the metaphor is apt and whether the central role of the physician is morally justified. To answer the question, we need to place the metaphor in context.

In Youngner's article, the ART metaphor is evoked in the same clinical setting, the ICU, as the WAR and TORTURE metaphors, and it is applied to the same phenomena to which the other two metaphors are applied. All the patients discussed by Youngner are critically or terminally ill; they are clinically unstable, frightened and depressed. Frequently, the patients whose deaths are being orchestrated are comatose, delirious and unable to communicate due to intubation. Yet, the ART metaphor gives a new meaning to these settings and phenomena, and construes the corresponding actions and events in a different way. The source domain of the new metaphor is art, and more particularly, the performance arts and music. The target domain is the process of dying and death. The ART metaphor highlights certain aspects of dying and medical care at the end of life, and hides others. As a result, the metaphor makes physicians' experiences coherent in the face of demands and obligations that appear contradictory and conflicting when we try to understand them through the lenses of the other two metaphors. By employing the concepts of art and orchestration, the metaphor highlights the creative character of caring for the dying and its spiritual dimensions. It acknowledges the individuality of the dying and the social context of the patient's death. This is expressed in the collaborative nature of orchestration and in the reference to the family and the health care team. All this seems to suggest that, at least on the face of it, the ART metaphor is patient-centered. Yet, in the conditions of intensive care, the patient's ability to exercise autonomy and decision-making power is restricted or altogether absent: "In short, the balance of power and control, if not moral authority, inevitably rests with the critical care physician and staff," concludes Youngner [23, p. 1617]. The language of "orchestration" implies that, in the treatment of critically and terminally ill patients, the physician, and, in a lesser degree, the health care team has significant freedom of interpretation. The physician has to exercise discretion in selecting the appropriate treatment for each patient, deciding how much information to present to the patient and the family, and how much involvement to encourage. The physician discretion entailed by the ART metaphor is supported by the considerable degree of uncertainty in medical decision-making and the highly individual character of clinical judgments. The discretion of physicians, medical uncertainty and the individuality of clinical cases give rise to the idea of improvisation in medicine and support the analogy between art and end of life care.

The ART metaphor gives a new meaning of the process of dying itself and transforms the way we think of death. In virtue of this, the metaphor not only organizes the experiences, thought and actions of physicians. It also provides a structure for thinking about and understanding the needs and experiences of the dying patients. The new understanding of dying and death expressed in the ART metaphor is based on the existing, culturally shared conceptions of art, musical composition and orchestration. In contemporary Western culture, orchestration is a special kind of activity: it is creative, cooperative, and, to a certain extent, unpredictable. When extended to death, dying and end-of-life care, the term "orchestration" implies that the physician has (some) control over death in virtue of the availability of and control over various means for timing death in the ICU. Youngner writes, "Even when death is imminent, we have interventions that can postpone it for hours, days, or weeks in almost every instance. This is especially true in the modern, technology-dense, intensive-care unit (ICU), where medicine and highly trained health professionals can monitor, support, or substitute for nearly every life function" [23, p. 1617].

The Normative Significance of the ART Metaphor

The ART metaphor incorporates death and dying into the sphere of medical care and re-directs the physician's attention and efforts to the process of dying. The metaphor is a search for a balance between the fight for life and the acceptance of death. Death is no longer viewed merely as the limit or failure of medicine. Accepting the patient's death no longer amounts to abandonment. By extending the meaning of art and orchestration to the process of dying and death, the ART metaphor re-conceptualizes these experiences and the intrinsic conflicts associated with them. The metaphor allows health care workers to see dying, death and the moment of letting go as part of a meaningful, creative and unique experience. In such a way, the metaphor transforms the traditional goals of medical care and the meaning of the daily activities of the health care team. Within the framework of the ART metaphor, the goal of medicine is to facilitate the patient's transition in the process of dying and to facilitate a good death. It suggests that care for the dying should no longer be considered futile, nor should the patient's death be viewed as a professional failure. The efforts of the physicians are justified and various forms of medical treatment are deemed appropriate as long as they lead to the desired end, the least worst death. In such a way, the metaphor sanctions non-treatment as a legitimate option and withdrawal of medical treatment as a morally appropriate medical means to achieving a good death. The quintessence of the ART metaphor is captured nicely in a remark by J. Billing, an internal medicine associate, who notes that "personal notions of comfort, of "death with dignity," of an appropriate way to die, guide the work. Prolongation of life may be a goal, but so may be the avoidance of measures which promote survival" [24, p. 35]. The ART metaphor thus endorses good ("optimal," or "appropriate") death as a proper goal of medicine. It urges physicians to regard the individual steps in the treatment of a dying patient as parts of a larger, meaningful process, and to see themselves as active participants in this process.

The collaborative nature of the ART metaphor expands the social context of medical care at the end of life to include the loved ones of the patient. This aspect of orchestration is highlighted by Robert Truog: "Thoughtful and compassionate clinicians often will orchestrate the death of a patient specifically to ensure that the patient's loved ones are left with 'the least worst' memories possible" [25, p. 25]. Another major normative effect of the ART metaphor is that it facilitates a change in physicians' attitudes to death and a movement away from the belief in the absolute value of life.

The major weakness of Youngner's ART metaphor is its portrayal of the role of the patient in the process of orchestration. On the surface of it, the metaphor appears patient-centered: it highlights the goal of allowing a good death, and underscores the individuality of end-of-life care. On a closer examination, however, it is clear that the patient remains passive and isolated from the decisions concerning his own life and death. "Ideally, the ultimate authority rests with the patient. However, the realities of critical care make realizing this ideal extremely difficult," argues Youngner [23, p. 1617]. While in theory, the metaphor recognizes the value of the patient's autonomy, it breaks down when faced with the constraints of care in the ICU. This suggests that the structure of intensive care does not allow for truly patient-centered care at the end of life and that the ART metaphor, as employed by Youngner, remains physician-centered.

It should be noted that the ART metaphor itself does not necessarily entail the physician-centered interpretation advanced by Youngner. Many physicians have been critical of the power and dominance of the physician in end-of-life care. As Steven Miles notes, "Death belongs to families and to patients, not to the medical profession. It is failure twice when the medical profession steals the death and fails to palliate. We should be helpful stagehands, not inept stage managers" [26, p. 29].

There are two other ways in which the ART metaphor "misses" reality and masks important facts about care for the dying. First, the language of orchestration implies a sustained, coordinated, collective effort to achieve the desired, pre-established goal. But studies show that clinical decision making at the end of life does not follow a controlled, reflective model but takes a "cascade" form [27]. The final decision is the result of a long and intense bargaining and negotiation between the physician, the family and, less frequently, the patient herself. The negotiation "occurs in the context of differential power relations among professionals and patients and families" [27, 252]. The process begins with the illusion of choice, which the physician presents to the family and ends with the sharing of moral responsibility for the patient's death with either the family or the patient himself.

Lastly, the concept of orchestration could be misleading because it implies harmony and coordination within the health care team, which rarely exist. The reality in large hospitals is different: the physicians rotate frequently, they do not have long-term relationship with their patients, and the medical team often is unaware of the patient's wishes and values. The discourse of art and orchestration creates the impression of continuity of care, cooperation and good rapport between physicians, patients and the patient's family. The metaphor does not challenge the institutional context of end-of-life care and the limitations this context imposes on achieving optimal care. The metaphor addresses the needs of the physician, the medical team and the family much more aptly than it does the needs and interests of the dying patient. The analysis of the ART metaphor suggests that, while the emergence of new metaphors of death and dying can contribute to the improvement of end-of-life care, metaphors alone are not sufficient to bring about a real change in the treatment of the dying. For a genuine change to occur, the displacement of conceptual metaphors must also prompt changes in the institutional context within which the metaphors are imbedded.

Conclusion

The moral and professional experience of critical care physicians is shaped by the dominant metaphors of death and dying, which operate in that context, namely, the WAR, TORTURE and ART metaphors. They structure physicians' attitudes towards death and dying, influence their reasoning and behaviour, and influence their decisions concerning the kind and quality of care for the critically ill. The metaphors reflect certain societal interests, cultural and moral values and prevailing professional assumptions. This is not to suggest that metaphor should be avoided in medical discourse as subjective and dispensable. On the contrary, it is my view that it is unavoidable to use metaphors when talking and deliberating about medical and moral maters in clinical context, and especially when end-of-life care is concerned. The medical metaphors of death are context-dependent and affect how physicians understand their tasks and responsibilities, how they categorize the patient's condition and how they choose between alternative modes of treatment. Moral dilemmas at the end of life are understood and resolved by the use and shift of metaphors of death and dying which metaphors reassign moral responsibility, shape treatment decisions and change physicians' attitudes to the dying and to the mission of medicine. The connection between metaphors, institutional context, and the treatment of the patient is not accidental but points to the mutual dependence between normative constraints, underlying values, and institutional commitments. For this reason, we ought to recognize the metaphors we use and to evaluate their normative connotations. One important practical implication of our conclusion is that any effort to change the medical attitudes to death and to improve the care for the dying presupposes an understanding of the medical metaphors of death and dying. Such changes and improvements would involve the development of new metaphors through the process of education and socialization of physicians and would require changes in the institutional structures, legislation and policies, in which the medical metaphors of death are imbedded.

References

[1] Aristotle. *The Rhetoric of Aristotle.* Trans. L. Cooper. Englewood Cliffs, N.J.: Prentice-Hall, 1960.
[2] J. Locke. *Essay Concerning Human Understanding.* Ed. A.C. Fraser. New York: Dover, 1959.
[3] G. Lakoff and M. Johnson. *Philosophy in the Flesh: The Embodied Mind and Its*

Challenge to Western Thought. Basic Books, 1999.
[4] G. Lakoff and M. Johnson. *Metaphors We Live By*. Chicago: University of Chicago Press, 2003.
[5] R. Gibbs. *The Cambridge Handbook of Metaphor and Thought*. New York: Cambridge University Press, 2008.
[6] A. Pascalev, *Medical Metaphors of Death and Dying: An Ethical Analysis*. PhD dissertation, Dept. of Phil., Bowling Green State Univ., Ohio, 2002.
[7] *Chamber English Dictionary*. Cambridge: Chambers, 1988.
[8] J. Soskice, J. *Metaphor and Religious Language*. Oxford: Clarendon Press, 1985.
[9] J. Childress and M. Siegler. "Metaphors and Models of Doctor-Patient Relationships: Their Implications for Autonomy." *Theoretical Medicine and Bioethics* 5 (1984).
[10] T. Engelhardt. *The Foundations of Bioethics*. New York: Oxford University Press, 1984.
[11] A. Nordgren. "Ethics and Imagination." *Theoretical Medicine and Bioethics* 19 (1998).
[12] Centers for Disease Control, "Worktable 309: Deaths by place of death, age, race, and sex: United States," *http://www.cdc.gov/nchs/data/dvs/Mortfinal2005_worktable_309.pdf 2005.* Accessed February 3, 2014.
[13] R. Zussamn. *Intensive Care: Medical Ethics and the Medical Profession*. Chicago: The University of Chicago Press, 1992.
[14] P. Aries. *Western Attitudes Toward Death: From the Middle Ages to the Present.* Baltimore: Johns Hopkins University Press, 1974.
[15] P. Aries. *The Hour of Our Death*. London: Penguin, 1981.
[16] I. Illich. *Medical Nemesis: The Expropriation of Health*. New York: Pantheon Books, 1976.
[17] R. Momeyer. *Confronting Death*. Indiana University Press, 1988.
[18] C. Herzlich and J. Pierret. *Illness and Self in Society*. Baltimore: Johns Hopkins University Press, 1987.
[19] A. Pascalev. "Medical Images of Death and Dying in the Intensive Care Unit." *Journal of Medical Humanities* 17 (1996).
[20] L. Thompson. "Structure of Critical Care: An Overview." In *Major Issues In Critical Care Medicine*. Ed. J. Parillo and S. Ayers. Baltimore: William and Wilkins, 1984.
[21] A. Relman. "Intensive Care Units: Who Needs Them?" *New England Journal of Medicine* 302 (1980).
[22] American Board of Internal Medicine, *Caring for the Dying: Identification and Promotion of Physician Competency Personal Narratives,* 1996.
[23] S. Youngner. "Orchestrating a Dignified Death in the Intensive Care Unit." *Clinical Chemistry* 36 (1990).

[24] J. Billings. "On Being a Reluctant Physician." *American Board of Internal Medicine. Caring for the Dying: Identification and Promotion of Physician Competency Personal Narratives* (1996).

[25] R. Truog. "Is It Time to Abandon Brain Death?" *Hastings Center Report* 27 (1997).

[26] S. Miles. "Stage Managing Death." *Caring for the Dying: Identificaton and Promotion of Physician Competency. Educational Resource Document* 29 (1996).

[27] J. Slomka. "The Negotiation of Death: Clinical Decision Making at the End of Life." *Social Sciences and Medicine* 35 (1992).

Philosophical-Ethical Alleviation of Perceptive Experience of Corporeality in Terminally Ill Sufferers

Florin Lobont, Ionut Mladin

Introduction. Corporeal integrity and psychological identity

Patients with terminal illnesses have unique physiological and psychological needs. According to the scientific literature, there has been a significantly increased interest in the psychological dimension of the *corporeal image* concept.[20] Significant research showed its importance for both female and male sufferers of all ages, under multiple aspects including health, shape, weight, and attractiveness. According to Moore et al.,

> The experience of the body/self is thus involved in multiple dimensions, such as sensory, affective, physiological, etc., making a contribution to distinct portions of the neurosignature. Put differently, a pattern-generating mechanism exists in the brain that is capable of sustaining an image of the body upon which sensory [tunes] are played [1, p. 377].

The literature from various fields dealing with numerous types of terminally ill, from cancer to MS, reveals numerous communication problems affecting their care: "studies suggest [the existence of serious] problems in clinician-patient communication because delivering news about the ... diagnosis, treatments, [etc.] can cause the dialogue to falter" [1, p. 368]. In children suffering of

[20] We owe the remarks about cancer sufferers' body image – quoted [1], [2], [3], [4], [6] and [26] – to the excellent survey and synthesis of significant international research in the field made by Aurelia C. Bărbat, and summarized in the paper entitled "Cancer patients' body image under *phronesis*: bettering wellbeing through wisdom" she co-wrote with F. Lobonţ for, and presented at, the National Conference of Bioethics, Cluj-Napoca, 13-16 September 2012. This paper was an improvement of a previous conference paper from 2011 authored by Aurelia C. Bărbat and Dorin Bărbat, entitled "Body image and body message by cancer patients. A pluridisciplinary perspective".

corporeal decaying, life-threatening, illnesses, their self-image is a construct based on material, cultural, physical, social, cognitive and emotional experiences of the so-called *corporeal transgression* [2]. They express themselves quite radically when speaking about their body and show significant levels of anxiety. Adolescents affected by such illnesses experience anxiety and (distressed) self-consciousness triggered by the negative reactions of others, including "lower self–worth, more social anxiety, and more negative body image perceptions, [despite of not being regarded] ... as less attractive by observers" [3, p. 43].

With particular respect to cancer in adolescents, important research highlights experiences in terms of a complex fear: an embodied fear perceived as a threat to the personal self; the experiencing of fear related to the physical body; the experiencing of existential fear; and the experiencing of fear related to the social self [4].

Like in sufferers of younger ages, in case of adults suffering of terminal illnesses the existential dimension of their suffering include the

> diminished sense of self; the prospect of death; increased feelings of alienation, feelings of burden and dependency, or even the loss of the previous foundations of meaning. Even survivors often have difficulties in sharing their experiences with people who have not experienced similar conditions [5, p. 6].

In all age groups the experience of loss (by surgical procedures), or the atrophy of body parts prove to affect sufferers' definition of self. For example, in case of women,

> a mastectomy represents an assault on femininity, beauty, attractiveness, and the perceptions of oneself as a 'natural' and 'complete' woman. Breast cancer patients' perceptions of cancer are negatively affected by the social construction of the disease, which in turn is heavily influenced and shaped by the lay literature and the media [6].

Again with respect to cancer sufferers, but of older ages, Thome et al. describe their images of their identities as directly related to the following perceptions of their corporeality: *diseased* body, *disobedient* body, *vulnerable* body, and *enduring* body. Due to the diminished available physical, psychological and mental capacities, the lived body transforms itself in a restricted body, the disease being represented by the aged sufferer as a "[…] disruption of the lived body rather than as dysfunction of the biological body, reflecting two different dimensions of illness described as lived experience and scientific conceptualization" [7].

Neurophysiological perceptive and representational aspects of body image distortion

In the following, we will focus on the role of perception in relation to the body on the mental level and its relationship with the representation. In D.M. Amstrong's view, alternative perception is an intrinsic process, a component of introspective consciousness. The direct perception of body events is immediately completed by a representation of perceptual content [8]. This is relevant in view of the fact that patients who suffer bodily changes as a result of medical interventions have a specific characteristic of body perception and representation in the new condition. Since introspective consciousness is triggered by changing body conditions, patients require adjustments to how they self-respond to biopsychic change.

It is well known that physiological disorders induce disruptive neuropsychological changes in body scheme. Thus, subjective awareness of corporeality includes, in various degrees, elements of perception and representation of the body parts configuration and their spatial arrangement:

> One can use vestibular, kinesthetic, tactile, and to a limited degree also visual perceptions for inferring the actual position and configuration of one's body. Even in the absence of distinct afferents from these channels, one has a basic 'feeling' of where one's body parts are [9, p. 11].

Thus, in counselling and therapy interventions on suffering patients, corporeal awareness is an essential dimension. Patients have a fundamental knowledge of corporeality following life experience, and this knowledge is well implemented, correlated, mentally and physiologically. The observation of neurological symptoms such as the "phantom limb" and various forms of agnosia and apraxia, results in the major difficulties of body integration through amputatory medical interventions of organs or members, without taking into account the spectrum of specific neurological disorders.

In patients with cancer, a perturbation of the appearance of the body occurs especially when adaptation to the body image is immediate, for example in the case of amputations. Some disturbances of body appearance can be minimized by surgical reconstruction interventions. Surgical operations on internal organs, although not having an effect on external body appearance, trigger a mental discomfort that can be manifested by lack of self-confidence, isolation, etc. Hair loss following chemotherapy, pallor, are associated with changes in external physiological appearance. All of these patients require support from the family group. Women who benefit from breast reconstruction following amputation show an improvement in the frequency of negative cognitions with regard to body changes suffered. At the opposite there are women with disturbed feelings of femininity and concerns about their perception by partners.

In addition to the necessary social support for cancer patients, the following is added: "People with decisions to make appearance-changing cancer treatment need support when assessing how their thoughts, beliefs, and feelings about appearance may influence these decisions [10, p. 335]." In cognitive behavioural therapy, significant results are obtained in the intervention on body image distortions, and these results legitimize the application of the method to a wide range of pathologies; strengths and vulnerabilities of the body image are assessed, the contexts responsible for negative emotions, the coping strategies, the attribution of physical appearance to a specific body scheme. During the interventions diaries are used, also mirror exposure, monitoring avoidance behaviours, assessment of subjective physical appearance, establishing strategies to control distortions, promoting positive body experiences, etc. [11] Genetic and evolutionary perspectives emphasize the role of context, concrete life experience in body image formation. In this way, the medical context is widened by a biopsychosocial vision, which determines the acceptance that the therapy for cancer patients, and not only in these cases, should also include interventions on body image distortions in terms of physical appearance and body functionality.

Corporeality is the most relevant dimension in interaction with the mind. Through our biological constitution, we have instrumental concepts about the parts of our body that belong to us. The body appears to be "something else." But the appearance of the corporeal at the subjective level "engages what might be called an evocative-integrative treatment of our embodied being-becoming, that is profoundly different from, and usually incompatible with, an instrumental treatment of the body [12, p. 94]." It is obvious that, for the most part, medical treatment favors the instrumental approach. Thus, body-mind therapy promotes a new interpretation of body distortions in specific or associated pathologies. Also, somatic psychology highlights the role of life experience in body image formation. The experience of corporeality becomes a way of interrogation within reflexive consciousness.

Undoubtedly, this vision aims to overcome neurological theories, which explain the mind in the body / the body in the mind strictly on the basis of psychophysical laws, and advocates a holistic, systemic embodiment. In reverse, the mind-body relationship appears as a continuous interaction on neurological bases. In understanding this interaction, we resort to Antonio Damasio's research on the body's cerebral mapping, the formation of neural maps following this process. It was arguably showed that, once formed, these perceptual neural maps persist even if certain body components are amputated and subsequently prosthetized. In other words,

> the brain can simulate, within somatosensing regions, certain body states, as if they were occurring; and because our perception

> of any body state is rooted in body maps of somatosensing regions, we perceive the body state as actually occurring even if it is not [13] (p. 102).

Consequently, the brain represents the body as it was before the medical interventions. This explanation gives us the opportunity to consider the impediments of patients' adaptation to the new biopsychic alteration conditions related, at least in part, to the neurology of mind-body interaction. In the extension, patients compare their own physical state to that of others. Thus, adaptation to the family group and to the social groups is generally altered because of a discrepancy that exists between the representation of the body before the medical intervention and the body image restructured after this intervention.

In terms of enactivism, corporeal perception is in direct relation to the environment, as it is formed by sensorimotor action, and cognition derives from this relationship, the representation being eliminated. Body information is accessible to the subject as a result of sensorimotor action. We can ask the following question: In the case of patients who undergo medical interventions through which corporeality is altered, can we talk about a sensorimotor perception as it was before? Clinical evidence may help to answer this question in the affirmative way. From an external point of view, the perception of corporeality is framed in a dynamic system of causes and effects that interpose in the mind-body-environment relation.

This type of cognition of corporeality also involves the process of representation. Also, the socio-economic status of the subjects influences the relational dynamics of the brain-mind-body and, ultimately, the subjective perception of corporeality. Starting from this kind of consideration, V. P. Taylor argues in favor of an assembled model of embodied mind through which the perception of corporeality is placed in intersubjectivity, the social brain being considered as inborn. The author relies on research on mirror-neurons and considers empathy to be an essentially human natural gift [14].

"Wise therapies:" philosophical and ethical interventions through *phronesis*

Apart from the century-old primacy held for centuries by religiosity,[21] other forms of spirituality are playing an important, albeit insufficiently recognized, role in how people affected with life-threatening illness cope, in order to regain an integrated existence. On the other hand, the lack or progressive loss of

[21]According to Carmen Bărbat, religiosity contains a classical psychological dimension with three subdivisions (to think, to feel, and to act), and an interactional dimension, emerging from its socio-historical environment" [15, p. 123].

religiosity, or its incipiency should not prevent sufferers from regaining at least some meaning and dignity.

In the treatment of body image distortions in cancer patients assessments and interventions are still at the beginning, which is why we believe that these people can benefit substantially from the integrative perspectives of philosophical counselling. In addition to the perceptual component, the attitudinal component should also be considered in body image assessment, as the values and personal beliefs of these patients are also fundamental in the representation of their self-image.

Our view on the potential usefulness of philosophical and ethical counselling in such cases is inspired by the Aristotelian distinction between the forms of wisdom or skill that correspond to the three main types of activity (*theoria, poeisis,* and *praxis*) namely *episteme* (intellect), *techne* (craftsmanship), and *phronesis* (judgement) that is, practical philosophy that underpins good conduct or virtuous living. We wish to drawn attention to the latter, whose value is judged mainly by the quality of contemplation it promotes [16].

Unlike medicine or psychology, this type of intellectual assistance is not a technique *(techne)*. Without being confined to a particular method, philosophical counselling applies reflective skills acquired during the study of philosophy in dialogue with clients. Unlike in case of academic philosophy, "reflective skills are applied to real life persons, the dialogue consists of spoken words, and the time frame is the now, the actual moment the dialogue takes place" [17, pp. 38-9]. This is a form of intellectual consultation quite distinct from psychological therapies.

Philosophical counselling aims at conceptual clarification and wisdom, without explanatory recourse to psychological patterns or mental health constructs. Nonetheless, it … may have beneficial psychological effects that can be broadly described as therapeutic [18, p. 112]. Philosophical counsellors or practitioners generally resist labelling their clients as intellectually "ill" or "dysfunctional," preferring instead to regard them as equal partners in the view-examining dialogue. Many of them find inspiration first of all in the Stoic and Epicurean philosophers, who focused on the use of philosophical thinking to minimize emotional disturbance and restore emotional tranquillity. An increased sense of wellbeing is a natural consequence of wise living, and it is an outcome well worth pursuing for its own sake. "People are not disturbed by things, but by the views which they take of things" (Epictetus).

By employing *phronesis*, counsellors and other helpers can help the clients appraise the force and dignity of their own values, understand their personal experiences and adapt to their existential situation and thereby buffer them against their emotional troubles and devastation.

In other words, philosophical counselling has broadly "therapeutic" goals in that it aims at alleviating anxiety and enhancing personal wellbeing [18, p.113], not in aspiring to substitute any form of psychiatric or psychotherapeutic practice. That is why philosophical skills need to be supplemented with training in psychological assessment and counselling. techniques. But the philosophical practitioners also advocate a two-way road in the relation between philosophy and psychotherapy: according to Tim LeBon, such philosophical techniques are complementary to other counselling methods which may be better suited to managing a client's potentially troubled reactions and should be welcomed by practitioners who already hold accredited qualifications in therapeutic counselling [19, p. 17], [20, p. vii].

Profoundly informed by philosophy, all "wise therapies" appeal constantly to philosophy's most important elements: its object, methods and systematic nature. Its main domains are metaphysics, theory of knowledge or epistemology, and ethics and theory of values. Among the most important metaphysical questions relevant for counselling are those about the meaning of life, based on the fundamental assumption (shared by logotherapists) that the will to meaning is the main driving force in humans.

In epistemology, the justification of beliefs is reflected in various therapies' attempts to solve practical questions about the ways in which clients' convictions, preferences and decisions can become more rational, making their existence more satisfactory. The theories of life's guiding values are reflected by the attempts to assess the value of counselling and in those of critical analysis and clarification of values that govern the life of clients. The philosophical ideas of right and wrong help counsellors in facilitating the decisions of clients confronted with ethical or decision risks; they are also employed by counsellors confronted with specific ethical dilemmas.

According to Tim LeBon, the utility of philosophy in counselling is proved by its therapeutic engagement via the five methods it had generated, namely critical thinking, conceptual analysis, phenomenology, thought experiments and creative thinking [19, p. 4].

Critical thinking represents the testing of arguments' resistance to critical investigation and the pertinence of the reasons of their acceptance.

Conceptual analysis is a way of clarifying the sense of the assertions made in a given context via language analysis, questing for definitions and making distinctions in order to eliminate confusions. In counselling it proves its applicability both in analyzing evaluative notions (such as that of autonomy), and in clarifying the clients' problems concerning the meaning of life, freedom, happiness, faith, death, etc.

Phenomenology, in its applicative-therapeutic employment, is a

> philosophy according to which the events and the objects are understood in terms of our immediate experience of them, as they appear to us ... The therapies based on phenomenology are focused rather on the client's experience about the phenomena

> than on presuppositions, speculations or inferences [21, p. 136].

Although theoretically one among other methods, phenomenology in this particular sense of "being with the client's material" and the so-called hermeneutic dialogue – that in this context refers in essence to the joint interpretation of that material by the counselor and the client – constitute the dynamic framework of virtually all philosophical methods applied in counselling.

Thought experiments are mental experiments meant to ground, enforce or, to the contrary, to invalidate some aspects of knowledge. In counselling they take the form of scenarios concerning the clients—built together with them—that allow the exploration of things really important to them. The more recently developed philosophy for children devised simplified philosophical techniques adapted to children and adolescents.

There are numerous examples of experiments through which clients suffering from terminal illnesses received significant help in restoring the meaning of their lives, understood and assumed the limitations of human condition, regained their dignity and inner peace through acknowledging and praising their personal values and capacity of living through their reflective-philosophical selves [22, pp. 313-4]. Other techniques included the assuming and acceptance of physical decline and its inferiority to the intrinsic value of (the philosophically examined) life; learning to intensify the cherishing of good memories and the present moment [23, Ch. 3, 12, 13]. Other experiments included the dualist (Cartesian) de-centring of the client's identity from their body to their mind-body complex with the mind as the epicenter of their selves [24].

In its turn, creative thinking (that includes among is methods brainstorming and lateral thinking) has been included recently in philosophy through practical ethics. Whereas critical thinking assesses arguments, creative thinking's role consists in retrieving the thoughts that grounded people's arguments or views, thus helping us to identify the „best problems" (that is, those that can be assumed and examined philosophically and can bring self-worth through self-knowledge) and their solutions [25, p. 29].

Philosophical counselling and medical diagnosis

A number of writers have emphasised that the medical profession has priority in interpreting significant deviations from standards of normal emotional function as symptoms of biological disorders. However, philosophical practitioners should avoid the converse view: When

> equipped with expertise in critical thinking and knowledge of sophisticated worldviews, [one] may be tempt[ed] ... to construe all emotional complaints as consequences of philosophical error

> ... As we learn more about physical influences on mood and rationality, the insistence on exclusively cognitivist explanations looks increasingly dogmatic … [D]ogmatic cognitivism betrays an ignorance of the trend in contemporary philosophy towards theories of embodiment, which emphasise the complex integration of mind and body [18, pp. 115-6].

Moreover, philosophical practitioners who have not been trained in psychological assessment and treatment are not equipped to detect all potential hazards. For example, when analyzing the cancer patients specialists should take into account the so called "chemo brain" syndrome (as side effect after chemotherapy). The symptoms include cognitive impairment, memory, concentration and thinking problems that affect the patient-therapist communication [26].

Conversely, the various therapists from the medical domains *stricto sensu* may not be equipped to deal with diseases entailed by somatic disorders, including at corporeal image level. Just as confronting clients with their logical anomalies (sometimes self-defensive delusions) or dismissing the existence of cognitive disorders may provoke distress and injure their sense of wellbeing, the abrupt handling of the truth by medical staff in communicating with cancer patients can dramatically deteriorate the latter's fragile self-regard and sense of meaning. In the Dutch practitioner Dries Boele's words, "[s]ometimes, [in such] cases … a period of denial must be respected so that the client has time to adjust their complex constellations of interrelated personal sensitivities to the new reality." These are stages in which "some type of non-philosophical therapy may be necessary …, [as in them] a person cannot lead his or her life in accordance with autonomous and critical thinking" [27, p. 46].

Obviously, in such cases the philosophical practitioners lacking relevant qualifications should avoid all non-philosophical interventions and refer such clients to the suitable professionals. This means that practitioners should possess enough knowledge to generally discern whether the client's problems are medical or not. However, the philosophical domain of expertise is not compatible with the imposing or contradicting clinical diagnoses. Yet, Brown writes, this does not mean that philosophical methods should not be employed as a technique to help people with recognized mental health disorders.

> Philosophy can expedite improvements in cognition and wellbeing even for psychiatric patients. But instead of accepting psychiatric patients for therapeutic work philosophical counsellors should be urging the adoption of philosophical training amongst the psychiatric profession [18, p. 118].

As already stated above, medical authority has become widely recognized as crucial in the diagnosis and treatment of various pathologies that involve distortions of corporeality. We want to emphasize that the philosophical counselling of the symptoms of bodily affections can be framed in the conceptualization of the experience of the human being as a whole. The patient as a human being interprets his mental and physical states in pathological conditions, depending on the social and cultural factors, life experience, that is why the counsellor can use the hermeneutical-phenomenological dialogue in order to improve his existential condition. For instance, James and Kevin Aho reached the conclusion that, alongside medicine, philosophical methods such as phenomenology can shape new horizons of understanding the specific condition of patients as human beings [28].

Conclusion

Terminally ill sufferers deeply need to have their concerns about their corporeal image addressed at self-perceptive level. The deterioration of this image contributes to their feeling of "disintegrated" existence." It is already a commonplace—albeit insufficiently considered—that supportive interactions are able to modulate the neuro-immune function in both healthy and ill people. As the group of severely ill patients / survivors becomes more and more recognized as a "distinct entity," more needs to be done in order to alleviate the effect of perceived appearance changes in all age groups and help them to develop a positive corporeal image. Philosophical and ethical counselling and other forms of dialogue therapy inspired by philosophy can contribute to these improvements.

Without crossing the boundaries of their competencies, philosophical practitioners can provide both patients and care staff with more means of coping with the existential suffering entailed by the life-threatening diseases. Alongside other forms of psycho-social and spiritual interventions, the millenary wisdom retained by philosophical practice can enhance the fighting spirit; reduce the feelings of helplessness; restore (at least in part) the meaning of life or finds inner resources of constructing new meanings, and thus decrease distress, anxiety and depression.

References

[1] R.J. Moore, R.M. Chamberlain, and F.R. Khuri. "A Voice that Wraps Around the Body – Communication problems in the Advanced stages of Non-small Cell Lung Cancer" *Yale Journal of Biology and Medicine* 74 (2002): 367-382.

[2] S.J. Williams and G.A. Bendelow. "Recalcitrant bodies? Children, Cancer and the Transgression of Corporeal Boundaries." *Health: an interdisciplinary Journal for the Social Study of Health, Illness and Medicine* 4(1) (2000): 51-71.

[3] J. Shroff Pendley, L.A. Dahlquist and Z. Dreyer. "Body image and Psychosocial Adjsutment in Adolescent Cancer Survivors." *Journal of Pediatric Psychology* 22(1) (1997): 29-43.

[4] A. Carlsson, A. Kihlgren and V. Sorlie. "Embodied suffering: experiences of fear in adolescent girls with cancer." *J Child Health Care, June 1* (2008): 129-143.

[5] A. C. Bărbat and D. Bărbat. "Body image and body message by cancer patients. A pluridisciplinary perspective," 2011, unpublished.

[6] F. Anagnostopoulos and M. Spyridoula. "Body Image of Greek Breast Cancer Patients Treated with Mastectomy or Breast Conserving Surgery." *J Clin Psychol Med Settings* 16 (2009): 311-321.

[7] B. Thome, B.A. Ebsensen, A.K. Dykes and I.R. Hallberg. "The meaning of having to live with cancer in old age." *European Journal of Cancer Care*, 13(5) (2004): 399-408.

[8] D. M. Amstrong. *The Mind-Body Problem. An Opinionated Introduction.* Colorado: Westview Press, 1999.

[9] M.J. Farah, T.E. Feinberg (ed.). *Patient Based Approaches to Cognitive Neuroscience.* Cambridge Mass: MIT Press, 2005.

[10] Th.F. Cash and L. Smolak (ed.). *Body Image. A Handbook of Science, Practice and Prevention.* New York: Guilford Press, 2005.

[11] J.L. Jarry and Th.F. Cash. "Cognitive – Behavioral Approaches to Body Image Change." 415-424. In *Body Image. A Handbook of Science, Practice and Prevention*, ed. Th. F. Cash and L. Smolak. New York: Guilford Press, 2005.

[12] B.B. Barratt. *The Emergence of Somatic Psychology and Bodymind Therapy.* London: Palgrave Macmillian, 2010.

[13] A. Damasio. *Self comes to mind. Constructing the conscious brain.* New York: Pantheon Books, 2010.

[14] V. Pitts-Taylor. *The Brain's Body. Neuroscience and Corporeal Politics.* Durham and London: Duke University Press, 2016.

[15] A.C. Bărbat. *Spiritualität, und Copingmechanismen bei Krebspatienten.* Timişoara: Editura Universităţii de Vest, 2011.

[16] Aristotle. *Nicomachean Ethics.* Oxford: Oxford University Press, 1978

[17] P. Harteloh. "On the Competence of Philosophical Counsellors." *Practical Philosophy* 10(1) (2010): 35-46.

[18] S. Brown. "The Therapeutic Status of Philosophical Counselling." *Practical Philosophy* 10(1) (2010): 111-120.

[19] T. LeBon. *Wise Therapy: Philosophy for Counsellors.* London, New York: Continuum, 2001.

[20] A. Howard. *Philosophy for Counselling and Psychotherapy.* Basinstoke: Palgrave, 2000.

[21] C. Feltham and W. Dryden. *Dictionary of Counselling*. London: Whurr Publishers, 1993.
[22] L. Marinoff. *Therapy for the Sane. How Philosophy Can Change Your Life*. New York: Bloomsbury, 2003.
[23] L. Marinoff. *Plato, Not Prozac!* New York: Quill, 1999.
[24] D.J. Chalmers. *Philosophy of Mind: Classical and Contemporary Readings*. Oxford: Oxford University Press, 2002.
[25] A. Weston. *A Practical Companion to Ethics*. New York: Oxford University Press, 1997.
[26] H.N. Kattlove and R.J. Winn. "Ongoing care of patients after primary treatment for their cancer." *CA Cancer J Clin.* 53(3) (2003): 172-196.
[27] D. Boele. "The Training of a Philosophical Counselor." 35–48. In *Essays on Philosophical Counselling*, ed. R. Lahav and M.V. Tillmanns. Lanham, Maryland: University Press of America, 1995.
[28] J. Aho and K. Aho. *A Phenomenology of Sickness, Disease, and Illness*. Lanham: Lexington Books, 2009.

PART V

BIOETHICS AND REGIONAL EDUCATION

Bioethics Education in Western Romania

Antoanela Naaji

Introduction

In the past decades, the study of bioethics has widely spread in medical schools across the United States. Most European universities also offer such courses, some of them on graduate and postgraduate levels. This is largely due to the increased complexity of new technologies and procedures used in biomedical research and healthcare. Issues such as the human genome, cloning, organ transplants, patenting of human tissue products, are only few examples.

Nowadays, biomedical researchers and physicians are more concerned with respecting the ethical principles which provide guidance in resolving ethical issues, while adapted to real situations. The codes and standards are those which provide a common set of principles and standards on which they build their professional and scientific work. However, researchers in the biomedical field and doctors should consider less visible ethical issues such as: ensuring patient self-determination and proper informed consent for medical procedures, end-of-life decision making, research ethics, reproductive medicine, and managed care and related economic issues [1]. For this reason, bioethics education for medical practice and research is essential and students and other professionals in the field need education on ethical issues and laws. Some arguments are: the legislation related to medical policies and patient rights is changing or is being updated very often; due to the achievements in biotechnologies and ICT field, healthcare systems function differently than before; clinical practice now involves decision-making on more complex and new issues, etc.

In Romania, Bioethics was introduced in the programmes of medical schools in the early 2000s. Being a new discipline, there are no comprehensive studies on its impact on medical practice and biomedical research.

The aim of this research is to evaluate how the utility and relevance of bioethics courses is perceived within specialized academic programs, and also as a branch of biomedical sciences, in western Romania. Considering the increase of bioethical education on international level, it is important to evaluate its weight within biomedical programmes and also to set the strategic directions and measures which have to be taken in order to improve bioethical education in Romania.

This study is presented as a necessity, due to the fact that bioethics is a fairly new subject in biomedical education and there are no related studies conducted in Romania. In the future, the study could be extended to the whole country.

For this purpose, a questionnaire was created and applied to 154 respondents, mainly students, teaching staff and biomedical specialists from 4 counties in western Romania. The questionnaire had 16 questions and a note which assured the respondents that the principle of confidentiality was respected even if the person could not be identified from the requested data.

Methodology

This section on methodology presents the method used for collecting data, the number of respondents, how the data was analysed, and what statistical tools were used.

The study was conducted using a questionnaire applied on the focus group made of 154 respondents from 4 counties in Western Romania. 63.6% of the respondents were students and 36.4% were specialists in the biomedical field, meaning residents, physicians, teachers, researchers, etc. in the biomedical field. They were divided by age in six ranges and five levels of education.

The questionnaire was applied in two forms: printed and through Google docs. The data extracted from printed forms were manually introduced in Google docs and in the end all data were collected in an Excel file. The statistical analysis of results was done using simple tools and also the SPSS program [2], [3]. All the results were discussed in accordance with the objectives of the research in order to evaluate the current situation of Bioethics education in western Romania and to extract some ideas of improvement and development in this field. The conclusions were extracted in accordance with the aim of the research.

Description of the questionnaire

The questionnaire was created based on international literature ([4], [5], [6]) and also considering the specificity of Romania [7]. It contains the following main parts: personal data, the situation of Bioethics studies completed by respondents, respondents' opinion on teaching Bioethics and the existing legislation. In total, there are 16 questions among which 11 are single-choice, 2 multiple-choice, 1 based on Likert-type scale, and 2 free-response questions. The addressed topics which have to be studied as part of Bioethics were also identified. The proposed questions were:

1. In which county are you studying/working?
2. Your age range is: *from 25 to over 65 years old* (6 options)
3. The last completed level of education is: *from high school to postdoctoral studies* (7 options)
4. Your current field of study/activity is: (3 options)
5. Your professional status is: (2 options)
6. Do you consider the study of Bioethics to be relevant to your activity in the biomedical field: *from extremely important to no opinion* (5 options)
7. During your training, did you study Bioethics? (2 options)
8. In which year of study do you think Bioethics should be taught: *from I to VI and every year* (7 options)
9. How many lectures or seminars do you think should be devoted to the subject of Bioethics? *From not at all to more than 4* (5 options)
10. Which of the following topics have you studied as part of Bioethics? (23 options)
11. Which of the topics below are important in the study of Bioethics? (23 options, scale from 1 to 5)
12. Which are, from your point of view, the most important advantages of studying Bioethics? List at least three advantages.
13. Do you face ethical challenges in your professional activity? (2 options)
14. Do you consider that what you have learned from studying Bioethics has been useful in your daily practice? (2 options)
15. Do you consider that bioethical issues are sufficiently covered by legislation in Romania? (3 options)
16. What are your suggestions on improving the legislative framework in the field of Bioethics?

Some of the questions refer to most important topics in the field of Bioethics, starting from philosophical justification and basic principles to more practical topics connected to real situations encountered in the healthcare system and n clinical research. The addressed topics were:

- basic principles of medical ethics;
- duty to provide care;
- fiduciary responsibility to patients;
- truth telling;
- informed consent;
- confidentiality;

- history of medical ethics;
- history of research ethics;
- philosophical justification of ethics;
- research ethics codes, declarations, and ethical guidelines regarding the ethical conduct of human subject research;
- justification of human participation in biomedical research;
- the ethics of clinical studies and trials design and conduct;
- the nature and scope of informed consent for trial participation;
- the concept of equipoise and its utility in the ethical justification of randomized and placebo-controlled, and standard care-controlled trials;
- issues regarding confidentiality and privacy in clinical research;
- inducements for research;
- populations in research ethics: the use of healthy and vulnerable (children, prisoners, mentally ill, etc.) subjects;
- research on biospecimens, foetuses, embryos and stem cells;
- use of animals in research;
- ethics of national and international public health;
- the organization and management of ethical committees or IRBs (institutional review boards);
- professionalism and the responsible conduct of research (the dissemination and publication of research results, intellectual property, etc.).

Statistical results and analysis

The data obtained from the questionnaire was collected in an excel file and the statistic calculations were done using simple statistic tools and the SPSS program. We measured simple distributions, different correlations, based on calculation of r and p coefficients, and cross-tabulations. Some of the presented results are: distribution by age, distribution by the last completed level of education, the relevance of studying Bioethics, the most studied subjects, the advantages of studying Bioethics, etc. Figures 1 and 2 show the results regarding the distribution of the focus group based on personal data (age) and the situation of Bioethics studies completed by respondents. Figures 3 - 5 and Tables 1 and 2 present the respondents' opinion about teaching Bioethics.

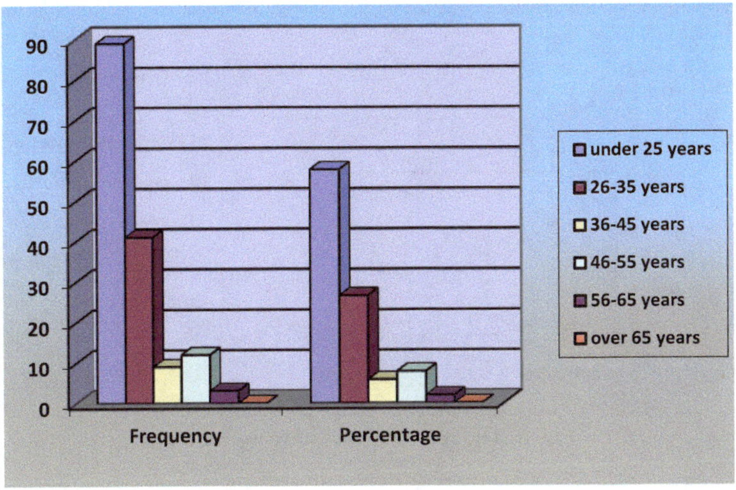

Fig.1. Distribution on age

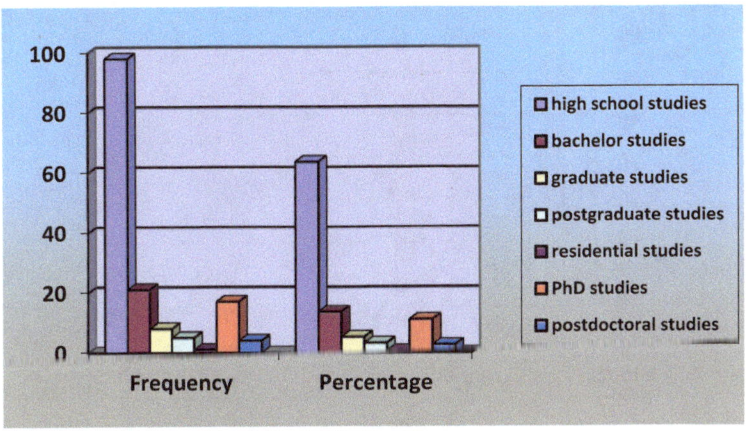

Fig. 2. Distribution on the last completed level of education

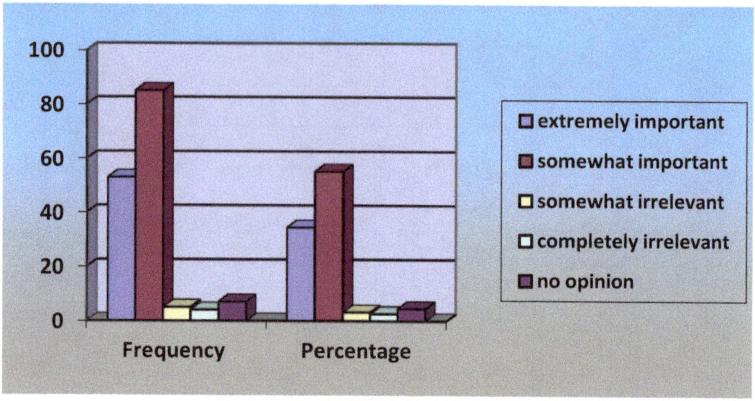

Fig. 3. The relevance of studying Bioethics

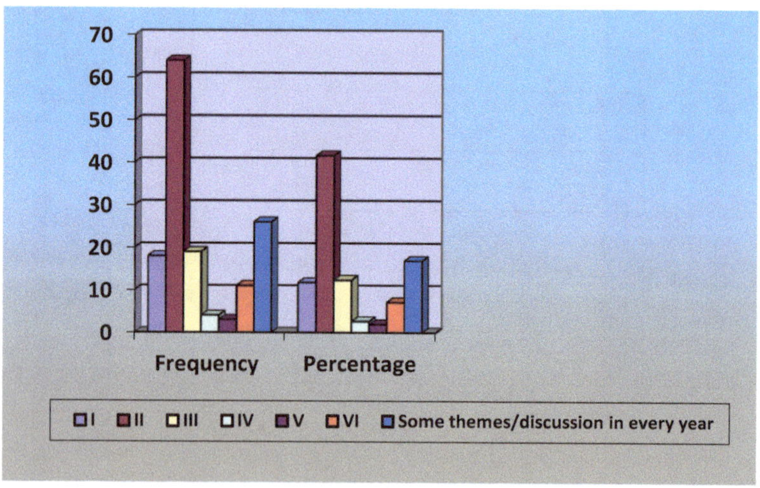

Fig. 4. The proper year to study Bioethics

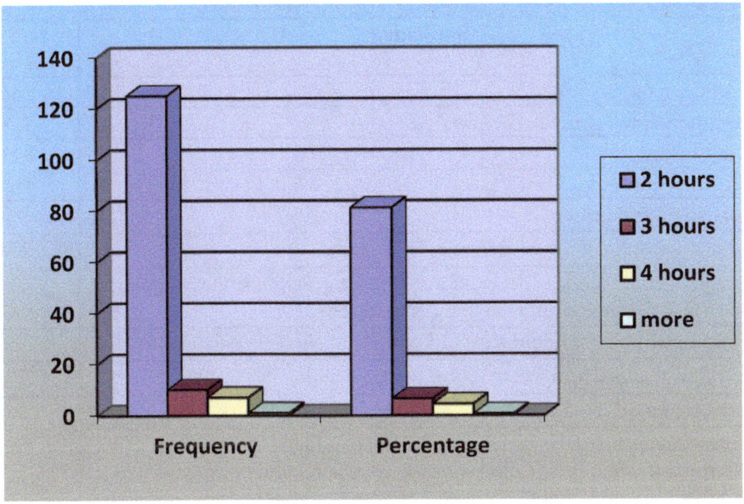

Fig. 5. Recommended number of hours/week

The results also show that 89.6% from subjects studied Bioethics and 85.7% faced ethical challenges in their professional activities.

The most studied subjects are presented in Table 1. The minimum points obtained were 32 points and the maximum 145. It can be noticed that there is a large difference between a group of subjects referring to day to day practice and another group which contains more theoretical or philosophical subjects.

Table 1. The most studied subjects

Subjects	No. of answers
Confidentiality	145
Basic principles of medical ethics	128
Truth telling	113
Informed consent	107
Duty to provide care	95
Fiduciary responsibility to patients	86
Research ethics codes, declarations, and ethical guidelines regarding the ethical conduct of human subject research	77
Issues regarding confidentiality and privacy in clinical research	75
Ethics of national and international public health	75
History of medical ethics	66
Professionalism and the responsible conduct of research (the dissemination and publication of research results, intellectual property, etc.)	64
The nature and scope of informed consent for trial participation	62
The ethics of clinical studies and trials design and conduct	59
Populations in research ethics: the use of healthy and vulnerable (children, prisoners, mentally ill, etc.) subjects	59
The concept of equipoise and its utility in the ethical justification of randomized and placebo-controlled, and standard care-controlled trials	53
Justification of human participation in biomedical research	52

Table 2 presents the respondents' opinion on the importance of the topics. The table presents only the topics considered most important (the mean/answer varies from 4.4416 to 3.0714).

Table 2. The importance of studied topics

The subject ordinated on importance	Mean/answer
Informed consent	4.4416
Confidentiality	4.4156
Truth telling	4.3636
Fiduciary responsibility to patients	4.3442
Duty to provide care	4.3182
Basic principles of medical ethics	4.2857
Professionalism and the responsible conduct of research (the dissemination and publication of research results, intellectual	4.2338

property, etc.)	
Issues regarding confidentiality and privacy in clinical research	4.1234
The nature and scope of informed consent for trial participation	4.0260
The ethics of clinical studies and trials design and conduct	3.9610
Ethics of national and international public health	3.9545
Research on biospecimens, fetuses, embryos and stem cells	3.9351
Justification of human participation in biomedical research	3.8896
The concept of equipoise and its utility in the ethical justification of randomized and placebo-controlled, and standard care-controlled trials	3.8896
Populations in research ethics: the use of healthy and vulnerable (children, prisoners, mentally ill, etc.) subjects	3.8896

The results also point out that 83.8% of respondents consider that studying Bioethics has been useful in daily practice, but only some of them identified these advantages in their answers. Some of the advantages of studying Bioethics, both for research and practice in respondents' opinion are:

- Research ethics:
  - Integrity of research (13 answers);
  - Confidentiality in research studies on human subjects (3 answers);
  - Engaging liability for consequences of medical acts performed on patients/participants in clinical studies (2 answers);
  - Ethics of utilizing human subjects in research (2 answers).
- Practice ethics:
  - Ensuring privacy (33 answers);
  - Truth telling (19 answers);
  - Informed consent (informed and assumed decisions) (16 answers);
  - Ethical approach to permanent medical acts (16 answers);
  - Improving of professional relations (14 answers);
  - Duty to provide care and help (9 answers).

It can be noticed that the respondents had opinions on this subject, especially related to bioethics in practice. This is normal because most of them are coming from clinics, universities (students and teaching stuff) and they are less connected with research.

The last part of the questionnaire refers to legislation (Figure 6). Only a small percentage of the respondents (14.9%) consider that the actual legislation is adequate and up to date.

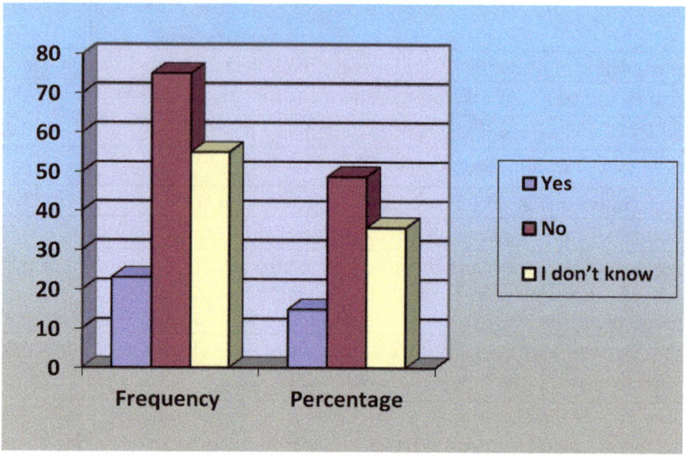

Fig. 6. Opinions about legislation in the field

Some suggestions for improvement are:

- Adopting guidelines and internal regulations on ethical aspects of biomedical research;
- Promoting integrity in research;
- Clear legislation on the act of donating one's body post mortem in order to be studied in higher education institutions;
- "I believe it is necessary to create a distinct branch of law: Medical Law. Integrating bioethics in the legislative system thus appears as a necessity."
- There should be a much more drastic law on verbal or other abuse perpetrated by patients against physicians and nurses;
- Clear and transparent laws implemented to promote ethics in medical universities.

Discussions

In this section, the results are analysed and some correlations are established in order to have a clear overview on the situation of studying Bioethics in Romania. Also, no statistically-significant relationships are mentioned. The following correlations were made:

The relationship between the age of the subjects and the studies in Bioethics (Table 3)

Table 3. The correlation between the age of subjects and the studies in Bioethics

Correlations			
		Your age range is:	During your training, did you study Bioethics?
Your age range is:	Pearson Correlation	1	.523**
	Sig. (2-tailed)		.000
	N	154	153
During your training, did you study Bioethics?	Pearson Correlation	.523**	1
	Sig. (2-tailed)	.000	
	N	153	153
**. Correlation is significant at the 0.01 level (2-tailed).			

We notice that there is a statistically-significant relationship between the age of subjects and the study of Bioethics ($r = 0.523$, $p = 0.000$), with younger subjects general studying more Bioethics. The strength of association between variables is $r^2 = 0.27$, indicating a strong effect.

The relationship between the age of subjects and the relevance of the study of Bioethics (Table 4)

Table 3. The correlation between the age of subjects and the relevance of the study of Bioethics

Correlations			
		Your age range is:	Do you consider the study of Bioethics to be relevant to your activity in the biomedical field?
Your age range is:	Pearson Correlation	1	-.177*
	Sig. (2-tailed)		.029
	N	154	154
Do you consider the study of Bioethics to be relevant to your activity in	Pearson Correlation	-.177*	1
	Sig. (2-tailed)	.029	

the biomedical field?	N	154	154
*. Correlation is significant at the 0.05 level (2-tailed).			

We notice that the relationship is statistically significant, since r = -0.177, p=0.029. Which presupposes that there is a relationship between the chronological age of research subjects and the relevance of studying Bioethics for the subjects' activity, namely, the older the subjects are, the more they believe the study of bioethics is relevant for their activity. However, the strength of association between the two variables is low, $r^2 = 0.03$.

The relationship between the level of studies and the study of Bioethics (Table 5)

Table 5. The correlation between the level of studies and the study in Bioethics

Correlations			
		The last completed level of education is:	During your training, did you study Bioethics?
The last completed level of education is:	Pearson Correlation	1	.446**
	Sig. (2-tailed)		.000
	N	154	153
During your training, did you study Bioethics?	Pearson Correlation	.446**	1
	Sig. (2-tailed)	.000	
	N	153	153
**. Correlation is significant at the 0.01 level (2-tailed).			

We notice that there is a statistically-significant relationship (r = 0.446, p = 0.00) between the level of studies and the study of Bioethics. The strength of association between the two variables is high, r2 = 0.19. Cross tabulation (Q3, Q7) shows that the tendency to study Bioethics pertains especially to undergraduate students, but also that from the total of 153 respondents, only 15 did not study bioethics as part of their professional training.

The relationship between professional status and the study of Bioethics (student/specialist comparison) (Table 6 a, b)

Table 6. The correlation between professional status and the study of Bioethics

a. Group Statistics					
	Your professional status is:	N	Mean	Std. Deviation	Std. Error Mean
During your training, did you study Bioethics?	student	111	1.0180	.13362	.01268
	specialist	42	1.3095	.46790	.07220

We notice that according to the data, there are statistically-significant differences between the study of Bioethics in students and in current specialists, t (151) = - 5.978, p = 0.00. Out of the averages, we see that current students studied Bioethics more than current specialists. Out of the averages expressed in Table 6.a. one can see that current students studied Bioethics more than current specialists. (As for the item "Did you study Bioethics as part of your training:" we rated responses 1=yes and 2=no).

b. Independent Samples Test										
		Levene's Test for Equality of Variances		t-test for Equality of Means						
		F	Sig.	t	df	Sig. (2-tailed)	Mean Diffe-rence	Std. Error Diffe-rence	95% Confidence Interval of the Difference	
									Lower	Upper
During your training, did you study Bioethics?	Equal variances assumed	226.114	.000	-5.978	151	.000	-.29151	.04876	-.38785	-.19516
	Equal variances not assumed			-3.977	43.554	.000	-.29151	.07330	-.43928	-.14373

The relationship between the current field of work and the benefits of studying Bioethics (Table 7)

Table 7. The correlation between the current field of work and the benefits of studying Bioethics

		Your current field of study/activity is:* Do you consider that the theoretical knowledge gained from the study of Bioethics bring substantial benefits in your daily practice: **Cross tabulation**				
		Count				
		Do you consider that the theoretical knowledge gained from the study of bioethics brings benefits in your daily process?				Total
		yes		no		
Your current field of study/activity is:	Medicine	84,5%	120	15,5%	22	142
	Biology	83,3%	5	16,7%	1	6
	Biomedical	80%	4	20%	1	5
Total			129		24	153

We notice that for most subjects theoretical knowledge obtained as a result of studying bioethics brings substantial benefits in their daily activity.

The relationship between professional status and the evaluation of legislative regulation of Bioethics in Romania (Table 8)

Table 8. The correlation between professional status and the evaluation of legislative regulations of Bioethics in Romania

Your professional status is: * Do you consider that bioethical issues are sufficiently covered by legislation in Romania: **Cross tabulation**					
Count					
		Do you consider that bioethical issues are sufficiently covered by legislation in Romania:			Total
		yes	no	I do not know	
Your professional status is:	student	17 (15.31%)	54 (48.64%)	40 (36.03%)	111
	specialist	6 (14.28%)	21 (50%)	15 (35.71%)	42
Total		23 (15.03%)	75 (49.01%)	55 (35.94%)	153

The table shows that, regardless of professional status, most subjects believe that aspects concerning Bioethics are not sufficiently regulated in Romanian legislation.

The analysis of the results established that there are no statistically-significant relationships between:

- the field of work and the study of Bioethics ($r = -0.031$, $p = 0.701$).
- the age of subjects and facing ethical challenges ($r = 0.50$, $p = 0.541$).
- the field of work and facing ethical challenges ($r = 0.80$, $p = 0.324$).
- professional status and facing ethical challenges ($r = 0.35$, $p = 0.662$).

Conclusions

Bioethics is a relatively new field in education, but universities have taken some steps to introduce this subject into their curricula, especially in Medicine (and to a lesser extent in Biology). Topics seen as more important are those that have more to do with the medical practice and less with research or theoretical issues.

Studying Bioethics is considered highly important by almost all respondents even if the existing legislation requires some improvements and actual implementation.

Considering the respondents' opinions on its importance, Bioethics seems to be a "promising subject" in Romania which needs to be taught in all universities. In order to provide medical care in an ethical way, students and specialists in biomedicine should be better educated on specific aspects of the ethical medical practice and research. The best ways to accomplish this goal is to extend the Bioethics education for medical students and resident physicians, and also through continuing education for practicing doctors and researchers.

References

[1] McCrary, S. Van. "The Role of Bioethics in Medical Education: A Crucial Profession Under Threat." In *American Institute of Biological Sciences*, 2001. http://www.actionbioscience.org/ biotechnology/mccrary.html. Accessed: January 15, 2018.

[2] Peacock, Janet and Philip Peacock. *Oxford Handbook of Medical Statistics*. New York: Oxford University Press, 2011.

[3] *** SPSS Tutorials, http://www.spss-tutorials.com/basics/. Accessed: January 15, 2018.

[4] Emanuel, Ezekiel J., Grady, Christine, Crouch, Robert A., Lie, Reidar K,, Miller, Franklin G., Wendler David (eds)., *Oxford Textbook of Clinical Research Ethics*, New York: Oxford University Press, 2008.

[5] Hulley, Stephen B., Cummings, Steven R., Warren, S. Browner, Grady, Deborah G., Newman, Thomas, B., *Designing Clinical Research*, Forth

Edition, Philadelphia: Wolters Kluwer Health/Lippincott Williams & Wilkins, 2013.
[6] Shamoo, Adil E., Resnik, David B., *Responsible Conduct of Research*, 2nd Edition, New York: Oxford University Press, 2009.
[7] M. Aluas, Maria, *Bioetică medicala [Medical Bioethics]*, Cluj-Napoca: Ed. Medicala Universitara I. Hatieganu 2016.

Acknowledgements

1. This research was realized with the support of the NIH Research Grant #R25 TW008171, Research Ethics Education in the Balkans and Black Sea Region, funded by the Fogarty International Center, U.S. National Institute of Health, PI: Prof. Rosamond Rhodes, Ph.D.
2. I would like to thank Lecturer Cristina Mercea, Ph.D. from Vasile Goldis Western University of Arad, Romania, for her assistance in the statistical processing of data.

(Bio)ethical Education for Young People in Macedonia

Dejan Donev

Introduction

Today, bioethics is a meaningful and complex field of thinking and activity. Although a young science, it is among the most important visions and concerns that humanity is facing. Every day new questions enter the scope of bioethical thinking -- issues and discussions above all. Meanwhile, many efforts are invested into debates whether bioethics should be primarily limited to the medical sphere (medicine, veterinary medicine, nursing, pharmacy, biological investigations on treating human lives) or whether it bioethics should be treated as a broader, large discipline, like modern practical ethics, in general. These questions, which result in scientific arguments, cannot be solved empirically – each one stands still in this debate for the moment! This is why good faith and collaboration among experts who have the duty to promote bioethics and its benefits are more than needed.

At the same time, knowing this, bioethical education cannot be postponed until the definite settlement of the scientific debate on the scientific purity of bioethics or the specific status of bioethics as a specialized or general science and practice (thus, whether the nature of bioethics is primarily biological or ethical). Many issues are strongly knocking on our doors: health problems, environmental pollution, violence and conflicts, human indolence, criminal deeds as well as negative behaviours of different kind. All these problems require serious and fast human action. This is why it is so important to introduce bioethical education to everyone.

In reality, the bioethical practice and science started at the same time as the development of ethical education, which included all young people (as well as older, if possible). It is so because the complex subject of bioethics matters not only to scientists and specialists, but they also concern every human being – all living as well as non-living beings on Earth. Therefore, everyone must be educated with regard to this new and elaborate area which is both spiritual and technical in nature. Furthermore, besides the numerous philosophical, social, and pedagogical issues it raises, the most important aspects in every educational system relate to the content and method of teaching. The aspect of content is a more interesting question for bioethical education as it is a new kind of education whose topics and arguments constructuct bioethics as a new science. It is therefore a field people are currently acting upon.

The development of the idea and practice of ethical and bioethical education in Macedonia

Starting with 1998, a campaign has been conducted for including ethical and bioethical education in the curriculum in the Republic of Macedonia. For at least 12 years, Prof. Kiril Temkov has been publishing in major journals and other media an open call to the Ministry of Education to persuade them to include first of all ethics and then bioethics as subjects matters on all levels of education -- or at least to permit the realization of the experimental ethical and bioethical education [1].

After years of struggle with the Ministry and the Bureau of Education, the process was accomplished, not only with the explanation of the meaning of these sciences, the essential need for them, the necessity of educating the young in these areas, the content of this kind of education, but also with the **practical pedagogical activity** in different schools at different educational levels – from kindergarten to master studies where bioethical education was mainly introduced as part of ethical education.

Extensive experimental teaching (courses) of ethics was introduced in high schools as well as in primary education and in kindergartens. Also, at the beginning of the 1990s, chapters on bioethics were introduced in the studies of Ethics and the History of Ethics for future philosophy teachers as well as in the education of other students from different faculties.

Experimental teaching in high schools and faculties

In 1998, there was an action of introducing ethics on all levels of education. The initiator of the process, Prof. Kiril Temkov, developed a program for experimental ethical and bioethical education in Macedonian high schools in four towns. The educational goals were to explore the possibilities of understanding and communication by young generations between ages 15 and 19 on ethical-bioethical problems and values. In two of them (specialized biotechnical schools), in the Agriculture-Veterinary High School in Drachevo, Skopje and in the Agriculture-Forestry School in Kavadarci, teachers Marija Taseva and dr. Jovan Lozanovski taught bioethics for the first time on high school level.

All questions of ethical education in these two schools had bioethical perspectives – understanding the position of human beings in the world and the proposed bioethical value system and moral norms. The program was composed of special lectures, such as lectures on the notion of bioethics, the history of bioethics, especially the ideas of bioethics, the system of human thinking, the evaluation and activity under bioethical dimensions, etc. This pedagogical experiment had a sense of bioethical education with general orientation as well as with contents of lessons, examples, and knowledge.

The high school in Drachevo founded a centre for researches in ethical and bioethical education for the youth, with the purpose of becoming an institute for this kind of education. However, the political changing of the school principals and the unfortunate discharge of useful spiritual and educative initiatives brought about a lack of continuity in the centre – initially, a special four-year programme for continuous bioethical education for high school pupils was established, but unfortunately, only a two-year course with one group was realized. The chance to implement bioethical education in high schools as well as use the extended experience in this field of science on other levels of education was stopped. At the same time, the organization of an experimental ethical teaching programme in the biggest biotechnical high school in Skopje was stopped by the Ministry of Education. Therefore, all the experience we had was reduced to only two years of education.

However, the former school principal, Jovan Lozanovski finished his dissertation on the methods of ethical education, as the first PhD on this matter in Macedonia. The results of the dissertation were published in Prof. Kiril Temkov's book *Good, Better, The Best* (Ethics and Youth) [2], where methods of ethical-bioethical education in high schools were explained and also the educational programmes for this kind of experimental education were analyzed, which later on became a groundwork for the programmes for ethical education in the IInd and in IIIrd classes (10th and 11th grades) in Gymnasiums in Macedonia.

On the level of higher education, all the activity done for launching the teaching of ethics in Macedonia was most important contribution in the process of propagating the idea for bioethics. First of all, in the education of teachers of philosophy there was an introduction on bioethical questions in their study of ethics and the history of ethics and then a special course on ecological ethics in graduate studies. A few years ago, another elective subject was given at the Faculty of Philosophy ("St.s Cyril and Methodius" University in Skopje) under the name "Foundations of Bioethics" [3]. Also, it was proposed that bioethics be introduced as a special subject in the master studies in Philosophy, an idea which was finally realized in 2012 [4], when I pushed the programme of bioethics through the Faculty of Law – a programme that was (and still is) an elective course for all students of this faculty's master programme.

Meanwhile, in 1998, the Faculty of Philosophy ("St.s Cyril and Methodius" University in Skopje) sent Prof. Temkov's letter to all faculties of the same university in a plea for the necessity of ethical education and for promoting bioethical research and courses, as well as for developing ethics in science). A few years later, it was proposed that the Faculty of Agriculture and Forestry introduced bioethics as subject of their basic studies, but the proposal was never followed through. At the same time, while bioethics became the main content of the new programmes in ethics, it was introduced [5] for students of Psychology, Pedagogy, Journalism, Arts, and other sciences.

In this context, one of the main aims of this struggle was to introduce bioethics in scientific development as graduate or postgraduate subjects in all fields, as part of the ethics of science (which should be obligatory in every kind of graduate or postgraduate studies). This project was presented at a round table discussion concerning this matter at the St. Cyril and Methodius University in Skopje in June 2007 (the resulting materials were edited under the title *Ethics in contemporary science*) [6]. Since then, bioethics represents 30% of the core content of the obligatory course titled "The ethics of scientific research" in doctoral schools [7].

Experimental teaching in primary schools and kindergartens

This entire process started with the idea that all children should study ethics and bioethics. The process was accomplished in 2000 with the experimental ethical education in sixth grade classes, where 12 year old children learned the basics of ethical knowledge, the most relevant moral values, as well as the special questions and norms of bioethics. This was the subject of the master thesis written by Ilinka Torbovska-Smilevska [8]. The lessons and texts used during this process were the well-known children reviews "Kolibri," published weekly in the oldest Macedonian newspaper "Nova Makedonija" between 2000 and 2002 (part of these texts were translated into Bulgarian and used in civic education in Bulgaria). A book with all this material was published under the title *Love and Care (Ethics for the Youngest)* [9], which discussed basic ethical and bioethical values of our time.

Further on, the most important experiment carried out was done in the second and third grades in the primary school of "Vlado Tasevski" in Skopje in the academic years of 2003/2004 and 2004/2005. The classes were taughts by Ljubica Topuzoska, graduated pedagogue, with an MA degree in philosophy. The two-year experiment conducted in her class represented a solid base for the master thesis titled *Ethical education in class teaching* [10] in 2004 and the doctoral dissertation titled "*Forms and methods of ethical education in class teaching*" [11] in 2006.

In teaching the second graders, the method of **Example and Edification** was primarily used. Over 80 ethical stories were created – all of them regularly published every Tuesday in the "Kolibri" reviews of "Nova Makedonia" (2003-2004); this way, the materials were available for reading and discussion to a large number of children in many schools. In the second semester, the content of ethical education was divided into six areas, two of them entirely bioethical - Health and Environmental care. Several years later, in 2007, these texts were published in a book titled *Ethics for Children* [12]. Also, the *Ecological lexicon* [13] was also used as additional material in teaching.

Third graders learned ethics in two forms - in the first semester the stress was on **Dilemmas** and **making a Choice**, while in the second one, **Moral Evaluation** was put in discussion. 20 special texts presented ethical and bioethical situation in which the heroes were children. The children were then requested to explain what they would do in particular cases. The bioethical themes from these stories were concerned with personal hygiene, the relationship with violent children, the problem of noise and powder rockets on New Year's Eve, etc. Later on, in the next semester, children were asked to give an ethical evaluation of somebody's behaviour. Here, bioethical themes were chosen from different forms of children's cultural and social behaviour (dance, help, etc.), as well as questions of the first love, of the relationship between brothers and sisters, meeting the strange world, etc. The children particularly enjoyed two of the texts. They were related to the problem of plant well-beingand the question of keeping pets and flowers received as gifts, as well as the responsibility one has for them as living beings.

In that time, bioethical content was represented in only one particular ethical-pedagogical material: the first ethical comic in Macedonia titled *Angela and Angel from the 1st A grade.* In the academic year 2004/2005, 19 comics were created and another 19 in 2005/2006 with interactive reading materials. The idea was to bring bioethical themes closer to children. In fact, the aim of the comic was to present the life of two neighbourhood friends who are going in school together in 1st grade. Most of the texts dealt with bioethical topics: personal, home, and school hygiene; regular nutrition; nonviolent relations between children; children's games in nature; the weather and the seasons; meeting and loving animals; comparing lifestyles in different environments (urban versus rural), etc.

Furthermore, the process had to be introduced to kindergarten, as well.. Between 2003 and 2005, 4 kindergartens in Skopje introduced ethical education on matters of Love, non-violence, peace, help and good behaviour. This involved the usage of many stories, designs, shows for children, education for teachers and parents, etc. A special game was prepared with eco-bioethical content on recognizing the home of animals, on plants and their fruits, on the ways they should be kept, on expressions of love for natural creatures, the child's behaviour towards nature, etc.. This process was recorded and a television documentary was produced on its information. The documents and ideas of the process were later published in the book titled *Ethics and Love* [14]. Because of this entire activity, some kindergartens today employ special Ecological education as part of their bioethical (human) upbringing.

Additional forms of spreading ethical and bioethical education for young people

Since the first experimental courses on ethics and bioethics were developed, numerous materials were published and disseminated. The materials of Prof. Temkov on experimental teaching are most noteworthy, but so are the numerous other publications: monographies, proceedings, articles, etc.

A series of lessons were published in a weekly newspaper as the first bioethical texts and lessons in Macedonia with the purpose of experimental ethical-bioethical education. Later, the series was a basis for the book *Good, Better, The Best (Ethics and Youth)*, the first edited book in large format published as private edition, translated into Serbian and after that edited in Macedonia in the same form [15]. In this book, Prof. Temkov published lessons on bioethics, ecological ethics, ethics of health and ethics of addictions which were used in the experimental teaching of Ethics in primary school – in the 6th (2000) and 2nd grades (2003/2004), while also providing value-related education for children aged 5-6 in kindergarten (2004).

There are also the books titled Love *and Care (Ethics for the Youngest)* and *Ethics for Children*, which discuss the definition of bioethics, the survey of development, the fields and values of bioethics, the moral norms in bioethics, and the future tendencies in bioethics.

In his book *Ethics today* [16], Temkov dedicated special parts to "Bioethics" and to "Addictions and Drugs" and he revealed how the status of ethics has changed – from the ethics of virtue to the ethics of nature. The book introduced and discussed the topics of bioethics, ecological ethics, health ethics, ethical aspects of infectious diseases and the ethics of addictions (especially of drugs). The book highlighted that today's the most important bioethical dimensions are the ones concerning the moral aspects of personal, social, domestic, economic and political ethics as values and tasks.

Other publications discussed similar topics, such as *Ethical Dictionary For The Youth* [17], *The Ecological Basic Textbook* [18], and many more, as well as new high school textbooks on ethical education: *Ethics for the 2nd grade* [19] and "*Ethics for the 3rd grade*" [20]. All these books for young people present bioethics as the most significant ethical problem of today. This is a new perspective, as bioethics mainly concentrates on specialized activities and the education of specialists for particular professions (medicine, biology, etc.), and it is "not part of wide base on comprehensive education with a mission to build a wide bioethical consciousness in everybody" [21, pp. 152-170].

The series "Ethicists" was first published in the newspaper "Start" (2003-2004) and later as a book titled "Ethicists – the 100 most important philosophers of moral: From Zarathustra to Bioethics" [22]. It presented the bioethical ideas of the most important ethical thinkers such as Albert Schweitzer, Pavao Vuk-Pavlovic, Rachelle Carson, Peter Singer, and others, from the first ethicists to the most modern ones.

In 2004, the Ethical Centre was founded in Skopje. The Centre's first seminar for students was dedicated to the problem of drugs, followed by other discussions of relevant bioethical issues.

The legacy of MA theses is also important: *Ethical education in class teaching* by Ljubica Topuzovska, *Ethical education in high schools in Macedonia* by Jovan Lozanovski [23], *Ethics as a subject in primary schools* by Ilinka Torbovska-Smilevska, to mention just a few. Doctoral dissertations should also be meantioned, such as *Forms and methods of ethical education in class teaching* by Ljubica Topuzoska, *Forms, methods and means in teaching ethics in high schools* by Jovan Lozanovski [24], which present the urgent need and forms of bioethical education as a part of the ethical education.

Finally, following the achievement of this level of discussions on bioethics education, besides the efforts of Prof. Temkov and his associates for opening the path for bioethics in the academic world in Macedonia, there is also the realization of the thematic block *Bioethics: Occurrence, Challenges and Opportunities* [25], edited by Dejan Donev in the scientific journal *Philosophy* from 2011 onwards. In addition, there the Proceedings of the first international interdisciplinary conference "Bioethics – The Sign of a New Era: Bioethics, Media, Law and Medicine" [26], that Donev organized with the Faculty of Law "Justinijan I" of the "St.s Cyril and Methodius" University in Skopje and the Centre for Integrative Bioethics from Kumanovo, in 2011, in Ohrid, Macedonia. Therefore, the ground was prepared for further actions of integrating Bioethics in the educational system of Macedonia.

Conclusion

The context of ethical education of young people in Macedonia is often dedicated to bioethical themes and problems. Young people learn bioethical questions very easily, especially children, who enjoy stories about animals and plants, and are interested in the bioethical side of existence. Hence, bioethical themes are best for introducing the study of ethics, as they also ennoble the ethical conscience with the idea of a good relationship with nature and other living beings. For instance, the experimental ethical teaching in the 2nd and 3rd grades in Skopje showed that bioethical content is unavoidable in the formation of a young person who will live in a new world of understanding, in good human relationships and proper behaviour towards nature. This experimental teaching programme showed that the children were very interested in bioethical issues and the bioethical themes easily found a path to their hearts.These researches used and analyzed the methods of ethical-bioethical education through edification, dilemma-choice and evaluation, as the best three methods of ethical education for young people. Furthermore, the method used for the ethical-bioethical education of young people was proven to be highly interactive and, at the same time, it discovered the "recipe" for the contemporary ethical education of the youngest children.

To conclude, the manifestation and theory of bioethics and is just as important as practical ethics. In the future, ethics might become bioethics, because the meaning of the biological factors of living are highyimportant (life, nature, the human being, as basic categories of bioethics), especially in era time when they are violated from various sides, including people's own negative behaviour. That is why we should engage in ethical education as a new kind of ethical-pedagogical activity. This is the type of education that should be implemented on all levels of education, so that society can understand and implement the new aspects of human living which are expressed through the reality of Bioethics.

References

[1] *Shortage of mercy for our young* ("Недостиг на милост за нашите млади"). http://star.vest.com.mk/default.asp?id=89470&idg=5&idb=1358&rubrika=Po%7Bta. Accessed: July 16, 2017.

[2] K. Temkov. *Good, Better, The Best (Ethics for youth) (Dobro, bolje, najbolje - Etika za mlade).* Niš: Samizdat, 2003.

[3] *Study program – philosophy* ("Студиска програма – филозофија"). http://www.fzf.ukim.edu.mk/ddtest21/public/uploads/files/00programinovi/Prilog_ br__4_studiska_rograma_-_filozofija.pdf. Accessed: July 16, 2017.

[4] http://ukim.edu.mk/mk_content.php?meni=163&glavno=40. Accessed: July 16, 2017.

[5] "List of subjects of the Faculty of Philosophy" („Листа на предмети од Филозофски факултет"). http://www.ukim.edu.mk/dokumenti_m/312_Filzoofski_lista-%D0%BD%D0%B0

%D1%98%D0%BD%D0%BE%D0%B2%D0%B0_web.pdf. Accessed: July 16, 2017.
[6] K. Temkov. *Ethics in contemporary science*. Skopje: UKIM, 2007.
[7] "List of proposed subjects" („Листа на предложени предмети"). http://www.ukim.edu.mk/dokumenti_m/262_Lista_na_predloznei_predmeti_genericki _znaenja.pdf. Accessed: July 16, 2017.
[8] I. Torbovska-Smilevska. *Ethics as a subject in prime schools* ("тиката како предмет во основните училишта). MA thesis, Dept. of Philosophy, St. Cyril and Methodius University. Skopje, 2001.
[9] K. Temkov. *Love and Care (Ethics for the Youngest) (Љубов и Грижа – Етика за најмладите)*. Skopje: Samizdat, 2002.
[10] Lj. Topuzovska. *Ethical education in class teaching* (Етичкото образование во одделенската настава). MA thesis, Dept. of Philosophy, St. Cyril and Methodius University. Skopje, 2004.
[11] Lj. Topuzovska. *Forms and methods of ethical education in class teaching* (Видови и методи на етичкото образование во одделенската настава). PhD dissertation, Dept. of Philosophy, St. Cyril and Methodius University. Skopje, 2006.
[12] K. Temkov. *Ethics for Children* (*Етика за деца*). Skopje: Samizdat, 2007.
[13] K. Temkov. *Ecological lexicon* (*Еколошки лексикон*). Skopje: Samizdat, 2003.
[14] K. Temkov. *Ethics and Love* (*Етика и љубов*). Skopje: Samizdat, 2007.
[15] K. Temkov. *How to be a good (Ethics for youth)* (*Како да се биде добар – Етика за млади*). Skopje: OHO, 2007.
[16] K. Temkov, *Ethics today*, (*Етиката денес*),Skopje: Epoha, 1999.
[17] K. Temkov, *Ethical dictionary for Youth*, (*Етички речник за млади*), Skopje: Epoha, 2001.
[18] K. Temkov. *Ecological basic textbook* (*Еколошки азбучник*). Skopje: Samizdat, 2003.
[19] K. Temkov. *Ethics for II class* (*Етика за II клас*). Skopje: Prosvetno delo, 2004.
[20] K. Temkov. *Ethics for III class* (*Етика за III клас*). Skopje: Prosvetno delo, 2004.
[21] K. Temkov. *Ethics* (*Етика*). Skopje: Epoha, pp. 152-170, 1998.
[22] K. Temkov. *Ethicians – 100 most important philosophers of moral: From Zarathustra to Bioethics* (*Етичари – 100 највлијателни филозофи на моралот: од Заратустра до Биоетиката*) http://www.arh.ukim.edu.mk/images/IzborniPredmeti/Estetika/ETICARI-PRV_DEL.pdf. Accessed: July 16, 2017.
[23] J. Lozanovski. *Ethical education in high schools in Macedonia* („Етичкото образование во средните училишта во Македонија"). MA thesis, Dept. of Philosophy, St. Cyril and Methodius University. Skopje, 1999.

[24] J. Lozanovski. *Forms, methods and means in teaching ethics in high schools* (Формите на работа, методите и средствата на наставата по етика во средните училишта). PhD dissertation, Dept. of Philosophy, St. Cyril and Methodius University. Skopje, 2001.
[25] D. Donev (ed.). "Bioethics: Occurrence, Challenges and Opportunities." *Philosophy* 31 (2011).
[26] D. Donev (ed.)."Bioethics – The sign of a new era: Bioethics, media, law and medicine." In *Proc. First International interdisciplinary conf.*, Skopje, 2012.

PART VI

LEGAL, POLITICAL, AND SOCIAL BIOETHICAL ISSUES

Ethical Concerns Regarding Mandatory Influenza Vaccination in Healthcare Practitioners

Miroslav Radenković, Ivana Lazarević,
Marko Stojanović, Tanja Jovanović

Introduction

An increasing global geographical mobility enhances the quickness infectious disease spreading over large areas. This is the highly significant reason why each country can assure immunization programs and related agendas which would be effective in order to reach and preserve the herd immunity [1]. Otherwise, when individuals are unvaccinated, the whole population absorbs the hazard [2]. In terms of preventive medicine, vaccinations have been confirmed to be the most successful pharmacological procedure. Nevertheless, soon after the initial development of vaccines, many ethical and legal questions appeared, regarding their necessity, actual effectiveness, beneficence, recipient autonomy, justice, and overall transparency.

The transmission of influenza to healthcare practitioners, as a result of nosocomial influenza outbreaks, has been well documented. This is not unexpected knowing that healthcare practitioners represent a heterogeneous group of professionals, including physicians, nurses, laboratory technicians, pharmacists, midwives, dental professionals, healthcare students, and administrative staff [3]. Consequently, healthcare facilities have placed wide-ranging efforts into large level prevention of influenza nosocomial transmission through vaccination programmes of healthcare personnel, with realistic long-standing prospects that vaccination of healthcare practitioners can result in the secondary protection of high-risk influenza patients. This is reasonable, since unvaccinated healthcare practitioners may significantly contribute to nosocomial influenza outbreaks in healthcare settings, too [4]. Moreover, the appearance of influenza viruses resistant to antiviral agents over the past time cycle additionally reinforces the role of vaccination as a head measure for the prevention of nosocomial influenza [5]. Nevertheless, regardless of well-aimed and continuous attempts of voluntary vaccination, unceasing high-level annual immunization rates are insufficient. Although vaccinations are the most effective pharmacological interventions for the protection of both healthcare workers and patients from infectious diseases which can be prevented by vaccines, one can always challenge the idea of mandatory influenza vaccination for healthcare practitioners, considering various ethical concerns [6].

Mandatory influenza vaccination in healthcare practitioners

Influenza vaccination is considered the most effective measure to prevent the transmission of the disease in healthcare facilities, which causes risks for both patients and healthcare practitioners [7]. However, vaccination coverage among healthcare practitioners continues to be suboptimal (<40% in Europe, >70% in the USA) [8, 9], even though vaccination of healthy healthcare practitioners is known to be associated with less influenza, less cumulative days of influenza-like sickness, as well as less cumulative days of absence from work [10, 11]. Since unvaccinated healthcare practitioners seem to be the main source of nosocomial influenza, according to the US Centers for Disease Control and Prevention, it is estimated that around 80% immunization coverage is necessary to prevent influenza transmission in healthcare facilities and to bring it to a safe level [12]. This is important given that hospital influenza has a particularly high mortality rate with a median of 16%, and as high as 60% in at-risk groups, such as transplant patients or patients in intensive care units [13]. Although it was established that under voluntary influenza immunization policy, hospital influenza outbreaks appear every flu season, most voluntary policies which augment vaccination rates in healthcare practitioners have not been very successful [14, 15]. Thus, it is clear that the effects of conventional educational programmes and campaigns generally only have a modest impact [16].

The objective of any mandatory vaccination course of action is to achieve prevalent vaccination among the eligible population [17]. This way, the primary goal of mandatory vaccination programmes preventing patients of higher risk of complications from becoming infected, suitably focus on decreasing morbidity and mortality [18]. Healthcare worker vaccination against influenza is also coherent with a collective professional duty to treat all patients impartially and to take essential safety measures as opposed to preventable harms [19]. Although mandatory vaccination was verified to function and accomplish more vaccination, it produces commotion in medical ethics for healthcare workers, as it was stated that a mandatory influenza vaccination policy infringes the individual's privilege to reject unwanted treatment, and furthermore underlined that any competent adult has the right to make a decision on their own treatment and refuse unwanted medical interventions [20]. However, this autonomy argument primarily focuses on individual rights of healthcare practitioners to refuse immunizations, disregarding the rights of patients [14]. In this case, one can always shift the focus from everyone else's rights against being infected with a transmittable disease to denial privileges. However, for supporters of mandatory vaccination, all scientific, ethical, legal, and financial requirements have been met, and accordingly, vaccination is consistent with a collective professional duty, and being immune is a part of the accountability of being a healer [21]. In other words, individuals who chose a healing vocation, unreservedly presume this duty to care and to make available the best possible medical treatment [22]. Widespread vaccination of healthcare workers against influenza should be considered standard patient care, while non-vaccination should be considered as maleficent care [23].

Looking for reasons regarding the decline of influenza vaccination, one may usually identify several groups of healthcare practitioners: unaware (consider themselves at a low risk), unbelieving (do not believe in the effectiveness), unmotivated (fear of side-effects and long-term complications), and unconcerned (vaccination stations are not available) [24]. Cognitive preconceptions involved in vaccination decision-making and potential obstacles which may alleviate their effect are numerous. For example, there is a tendency to prefer a potentially-harmful inaction to a potentially less harmful act, a tendency to prefer a known risk (no treatment) to unknown risk, a tendency to place more weight on costs and benefits today than on those realized in the future, a tendency to judge occurrence of side effects as likely and frequent, or a tendency to prefer natural products or substances even when they are identical or worse than synthetic alternatives [24]. However, considering that the common side effects of influenza vaccination are mild and self-limiting, and because the serious and extensive side effects of influenza vaccination are rare, if balanced with vaccination benefits, influenza vaccination must be considered safe [23]. To continue, overall efforts should extensively focus on expanding awareness and knowledge on advocated vaccines and their efficacy and safety [25].

Ethical considerations from both the clinical and public health perspectives have been used to investigate whether it is ethically acceptable to request the seasonal influenza vaccine for healthcare practitioners. Taking into account different components of justice, some argue that healthcare practitioner immunization exemptions should be limited to medical contraindications only [26], since justice allows for individual rights and freedom, but importantly, it limits autonomy when it harms or intrudes upon that of others. Thus, promoting the healthcare practitioners influenza vaccination should cover various activities including the distribution of written and electronic announcements promoting influenza vaccination, on-site vaccination within clinic with mobile vaccination availability, free of charge vaccination, distribution of educational materials, electronic distribution of information about the progress of the vaccination, and when applicable, mandatory completion of declination vaccination forms regarding medical and non-medical reasons [7, 9]. Despite this, the implementation of vaccination would not be full, even in the case of mandatory vaccinations, since healthcare practitioners may refuse it for personal beliefs, and simply be transferred to positions without contact with patients [27].

If voluntary acceptance of vaccination by healthcare practitioners is not optimal, the patients' welfare, public health and also the healthcare practitioners' own health interests should prevail over concerns regarding personal autonomy, and then fair mandatory vaccination policies for healthcare practitioners might be satisfactory [21]. In that way, mandatory vaccination policies must be carefully adjusted to adequately balance the risks and benefits of influenza vaccination for healthcare practitioners and patients [4]. In addition, the most vulnerable groups in our societies must be protected, such as the immunocompromised persons, the elderly (patients >65 years old), critically ill patients and young children in whom severe complications may develop, thus leading to hospitalization and even death [25, 28, 29]. Therefore, mandatory vaccination programmes need to be implemented through the widest cooperation of all available professional and non-profit organizations, as well as state, healthcare, and public health entities. This way, building trust in the healthcare system and reinforcing the importance of vaccination will be a feasible objective. After all, the public has the right to anticipate that healthcare practitioners and the institutions in which they work will take all the needed and sensible safety measures to keep them safe and minimize harm [19].

Conclusions

When public health is put at risk, healthcare associations and public health establishment have an obligation to take action, especially if an inexpensive, as well as effective method to accomplish patients' safety is available. Accordingly, the mandatory vaccination of medical workers may be an ethically-justified approach in cases in which the harm to patients and the overall population is considered to outweigh the autonomy of the individual healthcare practitioner. Moreover, mandatory vaccination of healthcare workers against influenza is coherent with a collective professional duty and it should be respected as a standard patient care. Promoting the influenza vaccination of healthcare practitioners should extensively focus on expanding awareness and knowledge about the efficacy and safety of vaccines. Mandatory vaccination programs need to be accomplished through the widespread collaboration of professional and non-profit organizations, as well as state, healthcare, and public health associations. Although there are many attempts to create workable and effective standards considering influenza vaccination for healthcare practitioners, many ethical challenges regarding mandatory vaccination are still very much present, and they require to be addressed.

Acknowledgements

The Ministry of Education and Science of the Republic of Serbia and the Fogarty International Center at the U.S. National Institutes of Health (NIH) supported this article through grants 175073 and 1R25TW008171-01A1, respectively.

References

[1] N.E. Moran, S. Gainotti and C. Petrini, "From compulsory to voluntary immunisation: Italy's National Vaccination Plan (2005-7) and the ethical and organizational challenges facing public health policy-makers across Europe," *J. Med. Ethics*, vol. 34, no. 9, pp. 669-674, Sep. 2008.

[2] C. Constable, N.R. Blank and A.L. Caplan, "Rising rates of vaccine exemptions: problems with current policy and more promising remedies," *Vaccine*, vol. 32, no. 16, pp. 1793-1797, Apr. 2014.

[3] P. Loulergue and O. Launay, "Vaccinations among medical and nursing students: coverage and opportunities," *Vaccine*, vol. 32, no. 38, pp. 4855-485, Aug. 2014.

[4] A.M. Stewart and M.A. Cox, "State law and influenza vaccination of health care personnel," *Vaccine*, vol. 31, no. 5, 827-832, Jan. 2013.

[5] H.C. Maltezou and S. Tsiodras, "Antiviral agents for influenza: molecular targets, concerns of resistance, and new treatment options," *Curr. Drug Targets*, vol. 10, no. 10, pp. 1041-1048, Oct. 2009.

[6] J.C. Tilburt, P.S. Mueller, A.L. Ottenberg, G.A. Poland and B.A. Koenig, "Facing the challenges of influenza in healthcare settings: the ethical rationale for mandatory seasonal influenza vaccination and its implications for future pandemics", *Vaccine*, vol. 26, Suppl. 4, pp. D27-D30, Sep. 2008.

[7] H. Guanche Gacell, A. Villanueva Arias, E. Guilarte García, R. Rubiera Jiménez and R. Nonato Alfonso, "A Successful Strategy for Improving the Influenza Immunization Rates of Health Care Workers without a Mandatory Policy", *Int. J. Occup. Environ. Med.*, vol. 6, no. 3, pp. 184-186, Jul. 2015.

[8] "Health Personnel Still Not Receiving Flu Vaccine", *JAMA*, vol. 312, no. 21, pp. 2203, 2014.
[9] H.C. Maltezou and A. Tsakris, "Vaccination of health-care workers against influenza: our obligation to protect patients," *Influenza Other Respir. Viruses*, vol. 5, no. 6, pp. 382-388, Nov. 2011.
[10] H.C. Maltezou, "Nosocomial influenza: new concepts and practice," *Curr. Opin. Infect. Dis.* vol. 21, no. 4, pp. 337-433, Aug. 2008.
[11] J.A. Wilde, J.A. McMillan, J. Serwint, J. Butta, M.A. O'Riordan and M.C. Steinhoff. "Effectiveness of influenza vaccine in health care professionals: a randomized trial," *JAMA*, vol. 281, no. 10, pp. 908-913, Mar. 1999.
[12] K. Quan, D.M. Tehrani, L. Dickey, E. Spiritus, D. Hizon, K. Heck, P. Samuelson, E. Kornhauser, R. Zeitany, S. Mancia, L. Thrupp, S.M. Tiso and S.S. Huang, "Voluntary to mandatory: evolution of strategies and attitudes toward influenza vaccination of healthcare personnel," *Infect. Control Hosp. Epidemiol.*, vol. 33, no. 1, pp. 63-70, Jan. 2012.
[13] C.D. Salgado, B.M. Farr, K.K. Hall and F.G. Hayden, "Influenza in the acute hospital setting," *Lancet Infect. Dis.*, vol. 2, no. 3, pp. 145-155, Mar. 2002.
[14] S. Wicker and H.F. Rabenau, "A Patient Safety Issue: Mandatory Influenza Vaccination for Health Care Workers," *Procedia in Vaccinology*, vol. 2, no. 1, pp. 101-105, 2010.
[15] S. McLennan and S. Wicker, "Reflections on the influenza vaccination of healthcare workers," *Vaccine*, vol. 28, no. 51, pp. 8061-8064, Nov. 2010.
[16] K.W. To, A. Lai, K.C. Lee, D. Koh and S.S. Lee, "Increasing the coverage of influenza vaccination in healthcare workers: review of challenges and solutions," *J. Hosp. Infect.*, vol. 94, no. 2, pp. 133-142, Oct. 2016.
[17] J. Flanigan, "A defense of compulsory vaccination," *HEC Forum*, vol. 26, no. 1, pp. 5-25, Mar. 2014.
[18] A.H. Antommaria, "An ethical analysis of mandatory influenza vaccination of health care personnel: implementing fairly and balancing benefits and burdens," *Am. J. Bioeth.*, vol. 13, no. 9, pp. 30-37, 2013.
[19] A.L. Ottenberg, J.T. Wu, G.A. Poland, R.M. Jacobson, B.A. Koenig and J.C. Tilburt, "Vaccinating health care workers against influenza: the ethical and legal rationale for a mandate," *Am. J. Public Health*, vol. 101, no. 2, pp. 212-216, Feb. 2011.
[20] A.C. Miller and D.W. Ross, "Mandated influenza vaccines and health care workers' autonomy," *Virtual Mentor*, vol. 12, no. 9, pp. 706-710, Sep. 2010.
[21] E. Galanakis, A. Jansen, P.L. Lopalco and J. Giesecke, "Ethics of mandatory vaccination for healthcare workers," *Euro Surveill.*, vol. 18, no. 45, pp. 20627, Nov. 2013.
[22] E.D. Pellegrino, "Toward a virtue-based normative ethics for the health

professions," *Kennedy Inst. Ethics J.*, vol. 5, no. 3, pp. 253-277, Sep. 1995.
[23] N. Cortes-Penfield, "Mandatory influenza vaccination for health care workers as the new standard of care: a matter of patient safety and nonmaleficent practice," *Am. J. Public. Health*, vol. 104, no. 11, pp. 2060-2065, Nov. 2014.
[24] A. Dubov and C. Phung, "Nudges or mandates? The ethics of mandatory flu vaccination," *Vaccine*, vol. 33, no. 22, pp. 2530-2535, May 2015.
[25] H. C. Maltezou and G.A. Poland, "Vaccination policies for healthcare workers in Europe," *Vaccine*, vol. 32, no. 38, pp. 4876-4880, Aug. 2014.
[26] L.M. Lee, "Adding justice to the clinical and public health ethics arguments for mandatory seasonal influenza immunisation for healthcare workers," *J. Med. Ethics*, vol. 41, no. 8, pp. 682-686, Aug. 2015.
[27] H.C. Maltezou, S. Wicker, M. Borg, U. Heininger, V. Puro, M. Theodoridou, and G.A. Poland, "Vaccination policies for health-care workers in acute health-care facilities in Europe," *Vaccine*, vol. 29, no. 51, pp. 9557-9562, Nov. 2011.
[28] G.A. Poland and R.M. Jacobson, "The age-old struggle against the antivaccinationists," *N. Engl. J. Med.*, vol. 364, no. 2, pp. 97-99, Jan. 2011.
[29] A.E. Fiore, T.M. Uyeki, K. Broder, L. Finelli, G.L. Euler, J.A. Singleton, J.K. Iskander, P.M. Wortley, D.K. Shay, J.S. Bresee, N.J. Cox and Centers for Disease Control and Prevention (CDC), "Prevention and control of influenza with vaccines: recommendations of the Advisory Committee on Immunization Practices (ACIP), 2010," *MMWR Recomm. Rep.*, vol. 59, no. RR-8, pp. 1-62, Aug. 2010.

Legal and Ethical Rules of Performing Medical Professions in Poland

Jakub Berezowski

Introduction

There is no definition of normative medical profession and medicine in Polish law. The concepts of health and its protection have not been normalized yet. In this respect, the definition of doctrine is still used, but it is justified by the great changing dynamics of these concepts. The absence of these definitions compensates for the relatively comprehensive definition of health benefits and health care benefits. The medical profession (in great simplification) means the provision of health services according to 2 sec. 1 point 10 of the Act of April 15, 2015 on medical activity (Journal of Laws, No. 112, item 654, as amended) and consists of actions aimed at the preservation, rescue, restoration, or improvement of health and other medical activities resulting from the treatment process or separate regulations governing their execution. At present, the medical professions, in addition to that of physiotherapist, the Polish legal system includes: doctors, dentists, nurses, midwives, medical rescuers, pharmacists, laboratory diagnosticians. This catalogue of occupations has consisted of main medical occupations, alongside ancillary professions, including technical occupations such as: dentist technician, analyst technician, radiologist technician, etc. [4, p.50]. The medical activity in Poland is regulated by numerous legal acts. On the one hand, the provision of health services requires appropriate legal forms for operators, and on the other hand, health service providers must prove their competence and qualifications.

Today, almost all medical professions in Poland are regulated professions, which results in the necessity of meeting the statutory prerequisites to obtain the confirmation of the right to practice. The necessity to regulate the right to practice is primarily due to the specific nature of the medical profession, which consists in the provision of health services. The provision of medical benefits needs to acquire relevant qualifications which will enable practice in relation to the patients. Performing medical professions are required to demonstrate the utmost care for the health and lives of patients. The treatment is usually a difficult and complicated process which causes the patient a lot of stress and suffering both mentally and physically. Patient care is a particular task which needs special normalization. The regulations should consider both the standards of competence and ethics. It is also necessary to set up regulators to define the requirements for providing health care and the necessary professional skills.

The Constitution of the Republic of Poland in Art. 17 sec. 1 (passed on April, 1997 by the National Assembly; OJ, No.7, pos. 4830) provides the possibility of creating a professional self-government, which would deal with the profession and its special requirements on behalf of the state. Custody refers to the correct, lawful performance of professional duties as well as to meeting the highest ethical requirements. The custody of profession, entrusted to an independent corporation, should be done in an extremely responsible way: on the one hand, it should maintain the independence of the professional self-government in relation to public authorities; and the other hand, it should be done in a way that would defend the public interest.

In carrying out the constitutional legislations, many professional groups entrusted the establishment of professional corporations. Over the years, the Polish Parliament enacted a number of laws appointing professional associations a number of professions, including medical professions. At the same time, independence is understood here as an independent intellectual process associated with the performance of professional activities involving decisions based on experience and qualifications of substance. It may be argued that the process of extracting professional corporations is a continuous process because every few years new regulations are written, despite the call for deregulation in the economic sphere. What is extremely important is that professional associations, including medical corporations have been authorized by the legislature to create a code of ethical standards binding its members. These directory standards form a substantially and universally binding law. On the one hand, there are the recipients of professional services, and on the other one, there is the basis for disciplinary responsibility, which is also involved in professional associations.

The road to medical professional corporations in Poland

The legal basis for the functioning of independent medical professions in Poland is contained in a few legal acts. It should be noted here that this is a profession which entitles provision of health care services on its own behalf in terms of normative competence. The history of professional self-reliance based on the corporate freedom of professional self-government was relatively long and varied in relation to different professions.

The corporation of doctors and dentists was the first to gain freedom and professional autonomy which was a natural process. Currently, this corporation operates under two acts: the Act of December 5, 1996, act on doctor and dentist professions (Journal of Laws No. 28, item 152, as amended), and the Act of December 2, act on medical chambers (Journal of Laws No. 219, item 1708, as amended). The norms of doctors were primitive to all medical professions. Two legal acts regulating the pursuit of the physician were enacted in post-partition Poland. The first law was the Act of December 2, 1921 concerning the pursuit of medical practice in the State of Poland (Journal of Laws of 1921, No. 105, item 752) and the second law was the Act of December 1, 1921 on the system and scope of medical chambers operation (Journal of Laws of 1921, No. 105, item 763).

In the case of nurses and midwives, corporate self-reliance in Poland is a relatively new phenomenon. The first comprehensive statute regulating these professions was passed on July 5, 1996 on the Nursing and Midwifery Professions Act (Journal of Laws of 2009, No. 151, item 1211, as amended). Currently, the Act of July 15 2011 on nursing midwifery occupations (Journal of Laws No. 174, item 1039) is in force, which has replaced the 1996 law.

The most recent corporation is the self-government of physiotherapists. The regulation of the professional competence of physiotherapists was postulated for many years. Finally, the passed bill was a parliamentary project in which a group of deputies recommended the regulation of the professional competence of physiotherapists. The rationale for the bill abounded in a number of arguments in favour of the special regulation of the conditions and principles of pursuit and thus the need to create a new professional corporation. Designers, above all, demonstrated that the physiotherapist profession is medical, which gives it a unique social character. There is no doubt that physiotherapists should show particular psychophysical characteristics and a high level of professional ethics. It is also important, as for every medical profession, to improve professional qualifications based on the latest medical knowledge and technology.

Until the Physiotherapist Act entered into force, physiotherapy activities were not regulated and physiotherapeutic activities could in fact be carried out by persons without professional qualifications, which unfortunately happened often. In practice, in order to pursue physiotherapeutic activities, it was enough to start a business activity and meet the conditions regulated by the Law of July 2, 2004 on the freedom of economic activity (OJ No. 101, item 1178, as amended).

The bill regulates in a complete way the profession of physiotherapist, as well as the rules of its execution. The construction of the bill is close to the standard of performing other medical professions such as those of doctors, dentists, nurses or midwives. What is important is that the legislator established the professional self-government in an unambiguous way in terms of the emergence of a new corporation -- not with regard to the interest of the self-government members, but also to the ethical standards affecting the regulation of competence standards and affecting the process of education.

The need for such solutions was questioned and will certainly continue to be debated. In the position of the Presidium of the General Medical Council, while questioning the need to give physiotherapists public trust, we read:

> The Presidium of the General Medical Council does not question the purpose of the statutory regulation of the physiotherapist profession. However, in the opinion of the Presidium of the General Medical Council, the project of recognition of the profession of physiotherapist as a profession of public trust and the establishment of professional self-governing people of this profession raises doubts. The profession of public trust consists of professions which perform tasks of a particular character from the point of view of public tasks and from the concern of pursuing the public interest [8].

In addition, this view highlights that the pursuit of public trust professions require the establishment of professional ethics, vows, corporate traditions, and it requires special professional and moral qualifications from its members. According to the Presidium of the General Medical Council, the position is in line with the ruling of the Constitutional Court of July 2, 2007 (file No. K 41/05).

The need for the regulation of the physiotherapist was also questioned on the same grounds on the project of the bill. The reviewer stated that there are no clear and valid arguments in favour of the creation of a professional self-government of physiotherapists, and that the current provisions clearly define the requirements which must be fulfilled for the pursuit of the profession of physiotherapist [9].

The scope of the competencies of medical corporations

The role of professional corporations in the Polish legal and health care systems remains significant. The scope of corporate competences is wide and the basic tasks of medical self-governments are:

- issuing decisions to confirm the right to pursuit the profession
- supervising the regularity of profession pursuit
- representing corporate interests to the executive and legislative branches
- raising the professional qualifications of corporate members
- passing codes of professional ethics
- engaging in professional responsibility
- organization and conduct of disciplinary judiciary

Such tasks require that the corporation actively and responsibly appoints functioning authorities, whose competence is regulated by law. In the case of the nursing and medical corporations, the authorities operate on both the central and the regional level. So far, the shape of the corporate body of physiotherapists leaves the activity of the authorities only on central level. However, the development of this self-governing body may result in the need of statutory changes and the appointment of regional authorities.

The need to distinguish professional medical corporations is justified primarily by the autonomy of the medical professions, as well as by the support given to the pursuit of a profession of public trust. The statutory definition of professional independence of physiotherapists is essentially connected to entrusting many functions to professional self-government. Such norms refer to professions of public trust which use a certain sphere of independence. These professions are also commonly referred to, in colloquial language, as "liberal" and benefit from legal elements of corporate autonomy. According to D. Karkowska, the characteristics of a liberal medical profession include, among other things, the protection of goods, intellectual independence, or individual relations with the patient [4, p. 34]. In my opinion, an essential feature of liberal professions is the real corporate independence. The objectives of the establishment and functioning of the corporation involve legal and organizational independence from public authorities, whose interference is limited to professional self-government, which is gaining regulatory, supervisory, judicial registration, and legislative functions. Although prerogative power is limited to corporate members, in practice they go much deeper. The impact of professional self-government in the legislative sphere is not overestimated. Strong corporate authorities successfully shape regulations not only for their own organization but also for the whole health care system.

For corporate self-government, it is also important to fulfil representative functions with respect to public authorities and society. This includes not only public administration but also local and regional authorities, as well as other professional corporations. The problem of negative corporate impact is extremely current, which could lead to the closure of corporations for new members. This is the place for state actions, which take care of the correct implementation of the constitutional custody mission for proper occupation by professional corporations.

A particularly important function of self-government is competence in the essence of normative character, which consists in shaping the principles of professional ethics. So far, the self-governments of doctors and dentists and the self-government nurses and midwives have set up their own ethical codes. In the near future, physiotherapists are expected to adopt the rules of professional deontology. Ethical standards not only determine the rules of the profession but also give disciplinary responsibility in the event of their violation.

Professional self-governments also gain the powers delegated by public authorities to become the same administrative bodies on behalf of the state. This dimension of self-government seems to be most important because it involves the transfer of trust to self-government bodies, whose decisions are binding not only for the members of the self-government but also directly to other persons and above all the recipients of the benefits provided by physiotherapists. J. Boć states that

> the division of powers between the organizational forms (organs) of the self-government and the organs of government administration that pursue supervision over the self-government or other organs of the public administration is a task of the legislator, and these competencies must be distinguished dichotomously [1, p. 48].

The main features of medical professions

Medical professions are regulated occupations. The definition of a regulated profession is readable in art. 5 Act of December 22, 2015 on the recognition of professional qualifications acquired in the Member States (Journal of Laws 2016, Item 65). It is defined as a set of professional activities whose performance is conditioned by the possession of formal qualifications required for the performance of these professional activities – this is required, subject to fulfilment of other conditions specified in these provisions. It should also be noted that regulated professions have been introduced in the Polish legal system by EU law and above all by the Directive 2005/36/ EC of the European Parliament and of the Council on September 7, 2005 on the recognition of professional qualifications. Since then, the European definition of regulated profession has been in force, 3 sec. 1 point of the directive. According to it, a regulated profession is a professional activity or a group of professional activities which, directly or indirectly, by virtue of laws, regulations or administrative provisions, have a special professional qualification; in particular, the use of a professional title is reserved solely under the laws, regulations, and administrative provisions for persons with appropriate professional qualifications.

The principle of professional independence according to D. Karkowska: "... is connected to the legal obligation of the 'personal' pursuit of the profession regardless of the legally permissible form of professional practice." [4, p. 259]. Independence as understood above refers primarily to the intellectual decision-making process which must provide medical services by persons who pursuit the medical profession. The decision-making process is primarily based on qualifications and professional experience. Professional independence does not mean that there are no normative limitations. These restrictions are included in a number of legal acts regulating the profession and above all in the commented law.

The definition of a liberal profession has not been normatively established in a way that it could provide a definition which will be universally applicable. Art. 17 sec. 1 of the Constitution of the Republic of Poland constructs only the concept of the profession of public trust, which should not be equated with a liberal profession. The current provisions apply to the concept of a liberal profession in four cases [3, p. 93]. First of all, this concept is defined by Art. 4 sec. 1 u.p.d. where a liberal occupation is a non-agricultural business carried out personally by doctors, dentists, veterinarians, dental technicians, fleeces, midwives, nurses, translators, and teachers in the provision of educational services consisting in giving lessons per hour if the activity is not carried out in favour of legal persons and organizational units without legal personality or for natural persons for the purposes of their non-agricultural economic activity, but for the personal pursuit of a liberal profession be carried out without employment on the basis of employment contracts, contracts of engagement, work contracts, and other contracts of a similar nature to people who carry out activities related to the essence of the profession concerned.

Such an approach to a liberal profession is primarily related the services of a freelancer. K. Wojtczak's doctrinal definition according to which

> the liberal profession is a personal and independent perfornce of an internally coherent set of intellectual, highly qualified (knowledge and practice) activities, systematically, in return for a selflessly determined fee, to provide benefits or services to clients, values of the general interest, in accordance with applicable legal norms, ethical and deontological principles [3, p. 95].

Taking into account these provisions related to the concept of liberal profession and its definition, all medical professions should be considered liberal occupations. These medical professions are of public trust as indicated by the nature of the constitutional regulations discussed and the statutory distinction of a corporate nature [7, p. 12]. K. Wojtczak states that the essential element of functioning is the existence of a professional corporation [6, p. 41].

Professional independence can be understood as intellectual freedom in the decision-making process of treatment and therapy. This autonomy may be constrained in the case of employee and organizational subordination in connection to the exercise of medical professions in medical entities and other entities in which health care is provided. However, the designation of statutory independence provides guarantees of independence in making decisions related to the proper performance of benefits in spite of obligations arising from a service or contractual relationship. This is justified by the commitment to pursuit of the profession with due diligence, in accordance to professional ethics, respect for the patient's rights, care for his or her safety, and the use of current medical knowledge. Such self-reliance should be guarded by a medical corporation, which, while controlling the pursuit of the profession, must pay attention to every manifestation of law violation.

The normative scope of professional duties

Competency catalogues of the members of medical corporations have mainly been defined in the laws regulating medical professions. In the case of doctors, the legal scope of their professional duties is presented in a relatively general way. According to art. 2 of the Act on the profession of doctor and dentist, performing the profession of doctor consists in granting by the person who has the required qualifications, confirmed by appropriate documents, health services, in particular: health examination, diagnosis and prevention, treatment and rehabilitation of patients, medical advice and also issuing opinions and medical certificates. As far as nurses are concerned, the law is wider than that of doctors. According to art. 4 of the Act on the profession of nurse and midwife, performing the profession of nurse consists of providing health care services, above all the followings:

- recognizing the patient's condition and health needs
- recognizing patient care issues
- planning and administering patient care
- self-giving in a given range of benefits: preventive, diagnostic, therapeutic, rehabilitation, and medical emergency services
- execution of medical orders in the process of diagnosis, treatment and rehabilitation
- making a decision on the type and scope of nursing care
- health education and health promotion

The best formulated was the physiotherapist catalogue, regulated in art. 4 sec. 2 on the physiotherapist's profession act. It is broadly formulated as following:

- functional diagnostics of the patient
- qualifying, planning and conducting physiotherapy
- qualifying, planning and conducting kinesiotherapy
- classifying, planning and conducting a massage
- commissioning of medical devices
- matching patients' medical devices
- teaching patients how to use medical devices
- health promotion

- issuing opinions and judgments on the functional status of persons undergoing physiotherapy and the course of the process physiotherapy
- teaching patients the compensation mechanisms and adaptation to the altered potential of body functions and activity

The pursuit of the medical profession on your own account involves the fulfilment of the requirements of the Healthcare Institutions Law. The most important requirements include:

- the obligation to make public the information on the extent and types of health services provided
- banning advertising
- the obligation to provide health care to a person who needs immediate care due to life threat or health issues
- the obligation to have rooms and facilities which meet the requirements of the general public, sanitary and installation standards appropriate for physiotherapeutic activities
- the obligation to use products conforming to the requirements of the Act on medical devices
- the obligation to ensure the provision of health services exclusively by persons performing medical profession and meeting the health requirements specified in separate regulations
- the obligation to sign a civil liability insurance contract covering damages arising of the provision of health care services or the unlawful abandonment of health care benefits

The prescribed competency requirements for performing medical professions are limited to medical issues related to the provision of health services. Running a business activity in the form of an individual practice involves additional normative obligations of a general nature which must be respected by all entrepreneurs.

Lastly, the normative obligations for the medical profession must be listed as defined by the Patient Rights Act of November, 6 2008 and the Patient Ombudsman (Journal of Laws 2009, No. 52, item 417). It is worth to emphasize that the rights of a patient regulated by a legal act largely correspond to the competences laid down in the medical professions law and, above all, the law on the profession of doctor and dentist. The most important rights of the patient are the main guidelines for medical practitioners and are defined in Art. 6 of the established law on right to health care benefits corresponding to the current medical knowledge, the right to the opinion of another doctor or nurse and the right to consult. In addition, Article 8 of the Act gives the patient the right to have health care services provided with due diligence by health care providers under conditions that meet specific requirements and the right to have benefits in accordance to the professional ethics of the representatives of the medical profession concerned. Therefore, the patient's rights also represent an obligation to medical practitioners to provide health care services. However, the enactment of the Patient Rights Act and the Patient Ombudsman was not particularly relevant for the legal status of patients because their rights are at the same time the duty of doctors, nurses and physiotherapists resulting from the presented laws and codes of medical ethics.

The postulated extension of the normative scope of medical procedures

The Act on Healthcare Institutions Law introduced in Art. 22 sec. 5 the optional possibility of introducing under the Health Minister's Ordinance the standards of medical treatment in selected fields of medicine or in certain entities performing medical activities. The motivation for normative standardization is, according to the mentioned recipe, the need to provide adequate quality of health services. So far, only low standardization has been decided on the provision of health care services. The following regulations were issued:

- The Ministry Of Health regulation of September 20, 2012 on standards of medical procedures for the provision of healthcare services in the context of prenatal care performed on a woman during physiological pregnancy, physiological childbirth, puerperium and neonatal care
- The Ministry Of Health regulation of December 20, 2012 on standards of medical treatment in the field of anaesthesiology and intensive care for practitioners
- The Ministry Of Health regulation of November 22, 2013 on guaranteed benefits for long-term care and nursing care

It seems appropriate to continue working on the introduction of further health care standards. Progressive standardization is necessary for the proper implementation of patient rights and proper provision of health care services. Standardization is certainly a method for improving the quality of medical services. Of course, it will not guarantee the elimination of so-called undesirable phenomena in the health care process, but will certainly reduce it. Moreover, if the undesirable phenomenon takes place, its evaluation will not depend only on the authoritative expert assessment but also on the implementation of normative procedures.

The introduction of postulated standardization will also have a desirable dimension for medical staff whose competence will be regulated in detail. Making decisions in the standardized health care process is less risky and therefore less responsive. Professional corporations should also play an important role in the process of standardization.

Conclusions

The process of regulation of health and medical law in Poland is continuous. It seems to still be far away from its full adjustment to the health system. There is the need to not only normalize the procedures but also to streamline the execution of the medical professions. The further development of the Polish health care system should be based on its actual substructure, which determines the actions of the members operating in the system. The range of health services must be specified as well as the competence of the persons responsible for the organization of the payment system according to constitutional requirements of equal access to health services. However, the services and the work of representatives in the medical professions seem to be settled, but, as noted in this chapter, management data is not perfect.

References

[1] J. Boć (ed.), *Konstytucje Rzeczypospolitej oraz komentarz do Konstytucji RP z 1997 r.* [Constitution of the Republic of Poland and a commentary to the Constitution of the Republic of Poland of 1997]. Wrocław, 1998.

[2] D. Karkowska. *Komentarz do ustawy o prawach pacjenta i Rzeczniku Praw Pacjenta. Komentarz* [Commentary on the Law on Patient Rights and the Patient Rights Ombudsman. Comment]. Warszawa, 2012.

[3] D. Karkowska. *Prawo medyczne dla pielęgniarek* [Medical law for nurses]. Warszawa, 2013.

[4] D. Karkowska. *Zawody medyczne* [Medical professions]. Warszawa, 2012.

[5] S. Poździoch. *Prawo zdrowia publicznego. Zarys problematyki* [Public health law. Outline of the problem]. Kraków, 2004.

[6] K. Wojtczak, *Zawód i jego prawna reglamentacja. Studium z zakresu materialnego prawa administracyjnego* [Profession and its legal regulation. A study in the field of substantive administrative law]. Poznań, 1999.

[7] A. Krasnowolski, *Zawody zaufania publicznego, zawody regulowane oraz wolne zawody. Geneza, funkcjonowanie i aktualne problemy.* Biuro Analiz i Dokumentacji. Zespół Analiz i Opracowań Tematycznych. Kancelaria Senatu [Public trusts, regulated professions and free professions. Genesis, function and current problems. Analyzes and Documentation Office. Team of Analyzes and Thematic Studies]. Senate Chancellery. Warszawa, 2013.

[8] Stanowisko Nr 6/15NII PREZYDIUM NACZELNEJ RADY LEKARSKIEJ z dnia 30 stycznia 2015 r. w sprawie projektu ustawy o

zawodzie fizjoterapeuty oraz o zmianie innych ustaw wraz z autopoprawką. [Statement on the draft act on the physiotherapist's profession and on the amendment of other laws together with the self-amendment]. Warszawa, 2015.

[9] Z. Snażyk, *Opinia prawna z dnia 3 lutego 2015 r. na temat projektu ustawy o zawodzie fizjoterapeuty oraz zmianie innych ustaw* [Legal Opinion of 3rd of February 2015 on the draft Act on the physiotherapist and other act]; druk sejmowy nr 3001. Warszawa, 2015.

A New Pro-Natalist Tool: Parenthood as a Form of Public Employment

János I. Tóth

The demographic situation in Eastern Europe

Today's rate of fertility, aging, and population decline is a new phenomenon. In over half of the world's countries, the fertility rate does not reach the ratio of 2.1 child/woman, necessary for population replacement, resulting in zero population growth [1]. The question that arises often is whether low fertility rate is good or bad [2]? The question is divisive. The league of experts I belong to consider too low (or too high) fertility rates to be bad because both are unsustainable [3].

Today, the global average fertility rate is 2.5 children per woman [4]. This high fertility rate average covers significant regional differences. According to fertility rate statistics, Africa (4.7 children/woman) and Europe (1.6 children/woman) are going through a serious demographical crisis [4]. While Africa suffers from the demographic consequences of a too high fertility ratio, Europe and Eastern Asia (1.6 children/woman) have the opposite problem. Therefore, it is misleading to only concentrate on the global average of fertility rate. In 1949, Sauvy argued that the notion of world population is just a statistical parameter and it is not filled with real content. From a demographic point of view, the world population consists of very different communities. Furthermore, specific demographic problems, be it overpopulation or underpopulation, can only be resolved by the affected population as a nation [5].

The terms "underpopulation," and "depopulation" refer to a problem that affects several countries in the developed world, such as Germany, Italy, Japan [6, 7]. The phrase "demographic winter" refers to what happens when human populations become unbalanced because the number of the children born is too low. The term "nuclear winter", popularized in the 1980s, alluded to the catastrophic environmental impact of a nuclear war – the long-term consequences of demographic winter could be equally devastating [8]. As Paul Demeny wrote:

> By the beginning of the twenty-first century, the average total fertility rate was 1.4 in Europe. Such a level, if maintained indefinitely, would result in a population loss of one-third from generation to generation, that is, roughly over each period of some 30 years. Some countries, notably in Southern, Central, and Eastern Europe, period fertility rates were at low levels without historical precedent for large populations [9, p. 306].

Globally, demographic winter is most severely felt in Eastern Europe because low fertility rates in the region are aggravated by high mortality and migration. Latvia used to have a population of 2.6 million in 1990, which shrank to under 2 million by 2014 and is expected to fall below 1.6 million by 2050. If the tendency continues, in 60 years the country is going to lose 40% of its population [10]. Hungarian society has been suffering from demographic crisis for decades: it is aging and its population is decreasing. Over the past 40 years, the country's population of 10.7 million has decreased by 1 million and the outlook is not encouraging either: without efficient intervention, the population will decrease to 7.9 million by 2060 and will further shrink to a mere 4.5 million of mostly elderly people by 2100 [11-13]. These changes will go hand in hand with diminishment in creativity, productivity, competitiveness, as well as reversed economic growth and decline in GDP, respectively, tax increases. Population decline and aging also entails the economic agony of the society.

The phenomenon justifies the state's vigorous intervention on behalf of a sustainable and stable population, without which no collective good is possible. The Hungarian government intends to resolve the demographic crisis afflicting the country by increasing birth numbers. Since the number of women in childbearing age is given, the only possible solution is to undertake more children, in professional terms, to increase the fertility rate. As Viktor Orbán emphasized

> It is important to highlight that the restoration of natural reproduction is a national cause; and it is not just one national cause among many, but the national cause. And it is also a European cause: not just one European cause among many, but the European cause [14].

Furthermore, he also declared that by 2030, Hungary must achieve the fertility rate of 2.1 children per woman, thus formulating a statement, respectively proposing a project that brings the problem of fertility rate to the forefront of politics [14].

What other solutions exist to avoid the demographic winter? There are two alternatives: one of them practical, the other one theoretic. The practical alternative is mass migration, the path chosen by Western Europe, while the theoretic solution is some sort of transformation of the human being (e.g. immortality) or of human reproduction (cloning, in vitro reproduction). Socially, as well as morally, both alternatives are much more problematic than striving to increase the fertility rate. Moreover, the traditional populations of Hungary and Eastern Europe are not keen on either mass migration or transhumanist modifications.

Consequently, for Eastern Europe, the best course of action for stabilizing its population is the radical increase in fertility rate. The solution will serve not only the survival of the nations, but will also promote well-being, as people who live in an aging and destabilized society already are and will be even more exposed to a great deal of suffering: rising taxes, decreasing pensions and subpar services. Setting the goal of achieving the fertility rate of 2.1 to ensure the stability of the population might seem to be an innovative and unusual objective, but it is also one that is socially necessary and ethically correct. As I have previously argued, this population policy objective must be set not only by Hungary, but by each country, including African countries. In a nutshell, each country, each political unit – nation, country, state – must strive to achieve replacement-level fertility [3].

The achievement of the objective raises questions about what pro-natalist tools are to be employed by governments. Bureaucratic pro-natalist tools (the ban on abortion and on contraception) violate reproductive freedom and are therefore morally unacceptable. The contradictions and the problems such as prohibitive policy measures can lead to high mortality, the increase in the number of orphans, and more abandoned children. They can be exemplified with Decree no. 770, implemented by the communist regime of Nicolae Ceauşescu as a tool to create a new and large Romanian population by restricting abortion and contraception [15].

Pro-natalist tools based on financial penalties (e.g. the implementation of a tax on childlessness) are also impracticable in democratic societies. Accordingly, only pro-natalist tools developed to reward and to stimulate child bearing can be regarded as realistic and morally acceptable. Hereinafter, I propose to focus exclusively on these types of pro-natalist tools.

The Two Forms of Natalism

Natalism (or pro-natalism) is the policy or practice of encouraging child bearing, especially by means of government support offered to promote higher birth-rate. Natalism, as a form of public policy, pursues the implementation of social and financial incentives to encourage the population to reproduce. One form of such financial encouragement is tax incentives to reward child bearing and child raising.

Other types of incentives used are the one-time baby bonus, child benefit provided for a certain period, tax reductions, paid maternity and paternity leave. The system in Sweden allows parents to share 16 months paid leave per child, the costs of the leave being supported in part by the employer and in part by the State. Other European countries, such as Italy, France, Poland, and Germany also offer incentives that combine monthly payments and bonuses. These measures are called *classical or 3-child natalist measures.* In what follows, I propose to point out that there is another possible form of natalist measures that can be called *radical or multi-child natalism.*

From a purely logical perspective, there are two ways to improve social statistical parameters: small changes undertaken by many or substantial changes undertaken by few. The same goes for fertility rate: either all fertile people undertake one more child or a few people undertake a significantly higher number of children. The first way is typical of classical pro-natalist family policies, while the second is the method proposed by radical natalist policy. The solution promoted by classical pro-natalist policies has yet to prove its efficiency, since the notion of "everyone should undertake one more baby" is not something that can be achieved by implementing economic incentives and it is impossible to enforce by means of bureaucratic measures either. Evidently, this form of pro-natalist family policy is also necessary, but the result, at best, is only that a few are going to have one more child. As these measures are far from being sufficiently efficient for Eastern Europe right now, radical pro-natalist policy measures are necessary.

I would like to point out that these two basically different pro-natalist family policies should not be confronted – they should rather treated as complementary tools that support one another mutually and are both employed to achieve the common goal: the 2.1 fertility rate.

Let us also examine the problem of low fertility from the perspective of content. In recent decades, in modern industrial economies, the participation of women in the formal work force expanded rapidly. This tendency, reflecting market forces and encouraged by government policy, is likely to continue. Many experts emphasize the *double burden* on women among the explanatory factors of low fertility, which includes child rearing and work outside the home. As higher birth numbers are socially desirable, the resulting policy task is to implement measures to make maternity and employment more compatible. The higher fertility in the European countries (particularly in Scandinavia) where such measures are heavily applied, compared with those countries (especially in Southern Europe) where they are largely absent, suggests that promoting compatibility (flexible working hours, liberal sick leave, through day care services and the like) are effective pro-natalist policies [9, p. 308]. As Demeny writes:

> But it is far from clear whether the fertility differential so generated is high enough to bring the total fertility rate back to replacement level. Steady labour force participation of women during the childbearing years can certainly be compatible with having one child or even two. It is likely to be far less compatible with sustaining, or even increasing, the proportion of women who have more than two children. Many career-oriented women voluntarily remain childless; many others prefer a single child. It follows that, to achieve average replacement-level fertility, the proportions of such women need to be counterbalanced by high enough proportions of women who have chosen third-, fourth- or even higher-order births. There is little indication at present that policies directed at enhanced compatibility achieve that result [9, p. 308].

In other words: the currently existing classical natalist measures encourage working women to undertake 2-3 children. As Demeny also indicated, in many regions, classical natalist measures are no longer sufficient. The reason for this is that the proportion of childless women and women who have a single child is too high. The proportion of childless women in Eastern Europe varies between 10 and 20% and their number is growing. The results of Miettinen's paper show that childlessness has increased in the age groups of 30–34 and 40–44 years among both men and women throughout Europe, with few exceptions. For example, female childlessness in the group aged 40–44 remains low (below or at 10%) in Hungary, Poland, Romania, moderate (11–15%) in France, Germany, the Slovak Republic and high (around 20%) in Austria, Italy, the Netherlands [16].

According to the data of the Hungarian Statistical Office (KSH), the proportion of childless women in Hungary in the 30-34, 35-39, 40-44 and 45-49 age groups in 2011 was as follows: 31%; 12%; 6% and 5% respectively. The degree of childlessness reaches the most dramatic levels in the 30-34 age group. Statistical figures improve for the higher age groups, but they are unlikely to return to the low value of the age group past fertile age. The findings of the Hungarian Statistical Office are consistent with the conclusion of Finnish researchers that the proportion of childless women is on the rise [17].

At this point, I propose to examine the group of 35-39-year-old women based on the number of the children they have: 0 children - 12%; 1 child - 19%; 2 children - 22%; 3 children - 8%; 4 children - 2%; 5 children - 0.5%; 6 children - 0.5%; 7 children - 0.02%. These figures illustrate that over 30% of the women in the examined age bracket have 0 or 1 child, while the ratio of the women who have 3-4 children is only 10% and only 1% of the women in the examined category have 5 to 7 children [17]. The proportion of women who have 0-1 child and their children pull the average downwards, while the women who have 3-6 children and their children pull the average upwards. Since the deficit in the number of children in the first category is significantly more substantial than the number of children on the other side, classical natalist policy measures can no longer be relied on to ensure the desired 2.1 children per woman ratio.

My thesis is as follows: classical (3-child) natalist policies can no longer achieve the fertility rate of 2.1 necessary for maintaining population stability in Eastern Europe. This statement is backed up by the rules of thumb of average calculation.

Let us now examine the transformations required in the above figures to achieve replacement-level fertility. For the sake of illustration, I propose not to refer to those who have 4 or more children and to round the fertility rate to be achieved to 2.0 children/woman – the two approximations more or less compensate each other. In this scenario, the number of women who have 3 children is equal with the number of those having one child, respectively twice the percentage of childless woman. Here I would also like to point out that half as many women with 6 children are needed as there are childless women. The rapport is illustrated in the following calculation:

1 child/woman + 3 children/woman = 4/2 = 2 children/woman.
0 child/woman + 2 x 3 children/woman = 6/3 = 2 children/woman
2 x 0 child/woman + 6 children/woman = 6/3 = 2 children/woman.

Taking the example of Hungary, the ideal proportion of women who have 3 children is 43% (=2x12%+19%), compared to the present figure of 8% in the group. An increase of this magnitude would be impossible to achieve by means of classical natalist measures, even if we assume that the increase in the number of women who have 3 children goes hand in hand with the decrease in the ratio of women who are childless or have a single child.

However, government initiatives to implement a radical natalist policy to encourage child bearing would be viable. As an outcome of these initiatives, the number of women with 3 children would narrowly balance the number of those who have just one child, while the number of women with 6 children would amply balance out the number of childless women. The fertility rate of 2 can be ensured if the number of women with 3 children in the full fertile age bracket (aged between 15-45) increases to 15% and the number of women with 6 children should reach 10%, this solution being more viable than increasing the number of women who have 3 children to 40%.

Consequently, the ideal, 2.1 fertility rate can be ensured only with a dramatic increase in the number of the women who undertake many (3-4) and very many (5-6) children. This also means that the classical (3-child) natalist policies also need to continue. With good reason, classical natalist measures rely on the hypothesis that 3, even 4 children can be raised by parents who are employed full-time, provided that the state offers comprehensive support to these families and society becomes more family-friendly (adopting a more positive attitude towards large families and household chores becoming more equitably shared by members of the family). However, raising 5-6 children becomes impossible with both parents working full-time. Consequently, such very large families need a completely new support system: the system of radical natalism. This raises a further question: what actual form should this type of support system take? As the principal objective proposed by classical natalism is to harmonize childbearing *and* employment for the mother, radical natalism, logically, needs to be elaborated on an entirely different foundation. The only way to create that foundation is to make society accept the rearing of very many children *as* a form of employment. To illustrate the principle with a formula:

Classical natalism:	raising 2-3 children + employment
Radical natalism:	raising 5-6 children = employment

Full-time Parenthood

The only way having a very large number of children can be encouraged is by getting society to accept this activity as a form of work that deserves payment. Let us call this type of reproductive activity *full-time parental work*. What I propose is that the state should elaborate a new form of public employment for parents *raising a high number* of children of their own (≥5) to provide the parents employed in this framework a wage calculated in proportion with the number of the children raised and *paid for the parent's lifetime*. The position should be available to young married couples or to married couples who already have children. The salary would be initially provided to the mother and it would be calculated in proportion with the number of the children reared and granted for the parent's lifetime. This salary would compensate for the high costs of child rearing and it would also ensure the time, the status and the necessary income for the activity [18].

A full-time mother who rears a large number of children apparently performs the same activities today as the mothers rearing the same number of children used to carry out in the past, but the status of the modern full-time mother is different. On the one hand, the implementation of the institution of full-time parenthood would allow women to choose between work and home-making. At present, women in Eastern European countries do not have this choice as the two-breadwinner family model forces them to work, that is, they are not allowed to be home-makers or they can stay at home as family care-givers only as low-status unemployed. Interestingly, for example, in the USA, wives have the freedom to choose between work and the status of being a homemaker. On the other hand, the proposed position is an option chosen freely, while in the past the role of the homemaker was traditionally imposed on women. Thirdly, the solution I propose would provide the woman payment for her work as a homemaker, which ensures her equal financial participation in the marriage.

Neither Hungarian, nor English language has the perfect term for this new job type. The proposed term, "parenthood as a form of public employment" implies only that the parent receives a wage-type remuneration in return for raising the children. My suggestion, however, would be that such parents receive life-long payments, even after the upbringing of the children is complete.

The concept of full-time parenthood raises numerous further questions – in what follows I propose to briefly discuss the most important issues.

How many children is a full-time parent required to raise?

The government budget would evidently encourage full-time parents to undertake the highest possible number of children, but the childbearing capacity of women is limited. These are the two aspects that need to be reconciled. I would like to point out that the greater the number of children required to obtain the status of full-time parent, the easier it becomes to adjust the institution to the present social structure. If families with two children could also qualify for the position, full-time parenthood would become a career option available for anyone, which would lead to the financial collapse of the state budget. If, by contrast, the position were made available only for parents who undertake 8 children, probably 6-8% of the population would be sufficient to deliver the expected results. However, having 8 children would excessively burden the participant families, therefore I suggest that the eligibility threshold be defined as having 5 children, with the system providing convincing incentives to encourage the participant families to have more than 5 children.

How many full-time parents does society need?

The answer to this question depends primarily on the population's average fertility rate and on the fertility rate achieved by full-time parents. For example, if the average fertility rate in a society is 1.3 and full-time parents have a statistical average of 5 children, the proportion of full-time women must be 21% to ensure the 2.1 rate. The following simple formula illustrates the scenario:

790 women x 1.3 children/woman + 210 women x 5 children/women
= 1027 + 1050 = 2077 children

In other words, 1000 women would have 2077 births – statistically speaking, one woman has 2.1 children. Columns 3, 5, and 7 in the table below illustrate the required ratio of full-time parents in a population with the average fertility rate ranging between 1.3 and 1.9 and the fertility rate of women employed as full-time parents varying between 4 and 6.

Table 1. Calculation of the ratio of full-time parents

Fertility rate of average women	**Fertility rate of full-time mothers (F4)**	**Rate of full-time mothers (R4)**	**F5**	**R5**	**F6**	**R6**
1.3	4 children/ women	29%	5	21%	6	17%
1.5	4 children/ women	24%	5	17%	6	13%
1.7	4 children/ women	17%	5	12%	6	9%
1.9	4 children/ women	9%	5	6%	6	5%

As the table illustrates, if the spontaneous average fertility rate in a society's female population is 1.3 (first row, first column) and the fertility rate of full-time mothers is 4 (second column), ideal fertility rates can be achieved if 29% of the population undertake full-time parenthood. If the spontaneous average fertility rate is 1.9 and the fertility rate of full-time mothers is 6, ideal fertility rates can be achieved with only 5% (last row, last column) of the population undertaking full-time parenthood. So, if the fertility rate of women who undertake full-time parenting reaches 6, the ratio of full-time mothers can stay below 20% of the total female population in the fertile age bracket.

Employment conditions

Full-time parenthood, as the term shows, is not supposed to be available on a universal basis (not as a subjective right), but as a form of employment. Unlike the baby bonus, child benefit payments or tax reduction, full-time parenting would be a form of public employment, similar to the employment structure of professional foster parents, who also need to fulfil very strict eligibility conditions. Candidates for the job must be selected by the employer - in this case, the municipality, the state. As the duration of the employment ideally lasts until the employee reaches retirement age, the process of determining individual eligibility must be more thorough and more rigorous than usual. The peculiarities of reproductive work also require the employer to proceed with extraordinary caution.

The selection of full-time parents can have two pitfalls. The eligibility conditions determined by the employer might be so strict that only a small number of the applicants can qualify. The other possible issue is that women or couples who neglect their children might also be approved. As a result, these unschooled, poorly educated children might become adults unable to properly integrate into the world and subsequently worsen, rather than improve the *real dependency ratio*. The overall dependency ratio is the number of workers relative to all non-workers, including the elderly, the young, the disabled, and those choosing not to work [19]. The selection of full-time parents must, therefore, be carried out by navigating between the two risks as if navigating between Scylla and Charybdis. The risk of having negligent mothers in the system is more severe as it would have irreversible consequences, therefore all the parents approved for the position must be able to raise children who can become at least average taxpayers.

I suggest the application of the following selection criteria for prospective full-time parent couples: selected applicants must be married couples, their housing conditions must be at least average, they need to have completed at least secondary education and they must have a tax record that proves reliability. Let us examine these conditions:

(i) Marriage is known to be the most stable form of relationship. Rearing many children is obviously challenging for the couple, therefore married status is a justifiable selection criteria.

(ii) Living among at least average housing conditions is essential to guarantee that the living conditions of children to be born are by no means inferior to the conditions enjoyed by their peers.

(iii) The requirement of minimum average level of education is important because the statistical data available for Eastern Europe infer that a child's school achievements are largely influenced by the level of their parents' education. The higher the individual's level of education, the more likely they are to be useful members of society – being employed, being a disciplined taxpayer, rearing children.

(iv) A taxpayer record that is at least average is important because reliable taxpayers are more likely to raise taxpaying citizens. The institution of parenthood as full-time employment incurs high social expenses that can be significantly offset by the taxes to be later paid by the children raised in the participant families. It is, therefore, essential to have the highest possible proportion of taxpayers among the children raised in these families. If full-time parenthood created crowds of people unable to work, people who rely on welfare benefits, the institution would aggravate, rather than mitigate the demographic crisis.

Full-time parenthood as a career path

The career of the full-time parent can be divided into the following phases: approval, the period of fertility, child rearing and child support, followed by retirement. The career of a full-time mother should ideally take the following path:

(i) She gets married at the age of 24, then the couple chooses the career of full-time parenthood for the woman and she is approved for the position.
(ii) In the fertile phase, the full-time mother gives birth to and rears 6 children. Obviously, during this period, only the woman is eligible for the position. Supposing that the woman gives birth to one child every 3 years on the average, the 6th child is born when the woman is 42 years old.
(iii) The educative period starts with the birth of the last child and it lasts, with gradually decreasing intensity, until the child becomes self-reliant. The first 12 years of this third phase, by the end of which the woman is 54 years of age, are more burdensome. During this period, the husband also becomes eligible for the position of full-time parent.
(iv) It would be unfair to send the older full-time parent back on the labour market at this age, therefore an indispensable income must be provided to the full-time parent (the mother or the husband). I suggest that the income should be calculated based on the tax revenues generated by grown-up offspring. Another aspect to consider is that by this phase, the couple's children will have become adults and if they have an average of 2 children of their own, the full-time parent will also participate in the raising of 12 grandchildren. If the full-time parents' own grandchildren live elsewhere, they can provide help to other full-time parents. The full-time parent in the phase (iii) and (iv) has the option to work in other areas and earn additional money.
(vi) Finally, the full-time parent reaches retirement age and retires.

The implementation of the institution of full-time parenthood requires a very lengthy, approximately 40 years long commitment not only from the individual, but from society as well. At the same time, the individual is always free to decide whether to continue being employed within this framework or to look for a job as a breadwinner for a large family. Evidently, the state cannot terminate the employment of the full-time parent without providing justification for the decision, but the recruitment of new candidates can be halted. This aspect allows both the parents and the state to implement modifications and corrections.

Social costs

The social costs of the institution of full-time parenthood are greatly influenced by the number of the full-time parents employed and their salary. The number of full-time parents, as previously discussed, would be 10-15% of the total female population in the fertile age bracket, that is, approximately equal with the number of primary school teachers, while their salary as public servants would equal the salary of a teacher employed in a kindergarten or in a primary school. The payment provided to full-time parents would increase fast, in parallel with the growing number of the children supported, reaching its peak when the woman is around the age of 42 and gradually decreasing afterwards.

When calculating social costs, the fact that a full-time mother can stay fertile for approximately 20 years must also be taken into consideration. This means that the number of full-time mothers in the system can be maintained stable only with an equal number of new entries. As a result, the costs of the process doubles after the first 20 years, but will not increase birth numbers. However, the number of full-time parents and the social costs of their employment will not increase further, because by the time "third-generation" full-time parents become employed, "first-generation" full-time parents retire.

During the first 15-20 years, the institution is going to be costlier, but after a while the system also starts generating revenue, as the children raised by full-time parents become taxpayers. The economic benefits driven by the institution (more employees, more entrepreneurs and more taxpayers) will manifest only if the children raised by full-time parents do not lag behind their peers during childhood or in adulthood, as I have previously mentioned.

Here I note that society is becoming more open to expanding the interpretation of the concept of work, which is in line with the spirit of the present proposal. In the Eastern European and East-Asian countries struggling with the crisis of demographic winter, the extra resources should not be wasted on providing *universal basic income* – much rather, these resources should be spent on raising fertility rates, and within that, allocated for financing full-time parenthood [20]. The advancement of Industry 4.0 is expected to release a substantial amount of industrial labour force, part of whom may consider full-time parenthood as a new career opportunity [21]. The way I see it, raising one's own children in Europe, or at least in Eastern Europe, will become an increasingly important way to earn a living. Otherwise, by the end of the century, the population of these countries will either vanish, or, more likely, their population will be replaced as a result of mass migration.

The implementation of the institution of full-time parenthood

The complete implementation of the institution of full-time parenthood obviously takes time. Let us calculate with the slow and cautious implementation of full-time parenthood and let us assume that 0.25% of the fertile-aged female population is provided full-time parental status annually. In an aging country with a population of 10 million and roughly 2 million women in fertile age, this means 5,000 full-time parenting positions. In this case, in 10 years' time, the number of available full-time parent positions will reach 50,000 and will generate a yearly increase of 17,000 in the number of infants born. In 20 years, the number of full-time parents will reach 100,000 and there will be 35,000 more new-borns. After this 20-year period, birth numbers through the institution of full-time parenthood stop increasing, but social costs continue to grow, reaching peak levels after 40 years. Thereafter, the costs associated with full-time parenthood will no longer increase. This way, the implementation of the system would take 40 years to fully develop, with the continuous employment of 100,000 full-time parents in the fertile age bracket and the same number of full-time parents older than that (> 45 years), giving birth to and raising 35,000 children per year. The institution of full-time parenthood will provide 385,000 births during the first 20 years, after which 700,000 children will be born over each subsequent 20-year period. The number of grandchildren, at a fertility rate of 2, would be the same, that is, 385,000 or 770,000, born to "average" families.

To conclude, the institution of full-time parenthood, complete with classical pro-natalist measures, could secure the fertility rate of 2.1 by 2035 and would thus be able to stop the decline and the aging of the population in Eastern European countries over the long term.

References

[1] United Nations, Department of Economic and Social Affairs, Population Division, "Figure 8: Population by Total Fertility (millions)", World Population Prospects, the 2010 Revision, 2011. http://web.archive.org/web/20121028012124/http:/esa.un.org/unpd/wpp/Analytical-Figures/htm/fig_8.htm. Accessed: August 18, 2017.

[2] A. Turner. "Is a Shrinking Population Always a Bad Thing?." *Social Europe* on 27 August 2014. https://www.socialeurope.eu/shrinking-population-always-bad-thing. Accessed: August 18, 2017.

[3] J. I. Tóth, "Demográfiai válság és a fenntarthatósági etika" [Demographic crisis and sustainability ethics]. *Valóság* 4 (2015): 26-33.

[4] Population Reference Bureau, "2016 World Population Date Sheet", http://www.prb.org/Publications/Datasheets/2016/2016-world-population-data-sheet.aspx. Accessed: August 18, 2017.

[5] A. Sauvy. " 'Le faux problème' de la population mondiale." *Population* 4(3): 447-462. English translation: "Alfred Sauvy on the World Population Problem: A View in 1949." *Population and Development Review* 16(4) (1990): 759-774.

[6] N. Folbre. "The Underpopulation Bomb." *Economix*. February 11, 2013 6:00 AM, http://economix.blogs.nytimes.com/2013/02/11/the-underpopulation-bomb/. Accessed: August 18, 2017.

[7] K. Kelly. "2013 What 'Should' We be Worried about? The Underpopulation Bomb." *Edge* 2013 https://edge.org/response-detail/23722. Accessed: August 18, 2017.

[8] S. E. Smoot. "*Demographic Winter the Decline of the Human Family.*" SRB Documentary film LLC, USA, min. 52, 2008. http://demographicwinter.com. Accessed: August 18, 2017.

[9] P. Demeny. "Population Policy." 294-313. In *Demography*, vol. II., ed. Z. Yi . EOLSS Publications, 2010.

[10] L. Mucenieks, the Head of the Organization of Large Families in Latvia, conference lecture, at the *11th World Congress of Families* (WCF) in Budapest, on 28 May 2017.

[11] B. Kapitány. "Demográfiai fogalomtár" [Demographic concept]. Budapest: KSH Népességtudományi Kutatóintézet, 2015. http://demografia.hu/hu/tudastar/fogalomtar/38-teljes-termekenysegi-aranyszam. Accessed: August 18, 2017.

[12] "Ha így folytatjuk, 2100-ra tíz magyarból négy marad" [If we continue like this, by 2100 only one Hungarian will remain out of ten]. *HVG,* 10 August 2015, http://hvg.hu/gazdasag/20150810_Ha_igy_folytatjuk_2100ra_lefelezodunk. Accessed: August 18, 2017.

[13] United Nations. "World Population Prospects 2017, Population Division" https://esa.un.org/unpd/wpp/Graphs/Probabilistic/POP/TOT/. Accessed: August 18, 2017.

[14] Prime Minister Viktor Orbán's opening speech at the 11th World Congress of Families (WCF) in Budapest, miniszterelnok.hu, 25 May 2017, Budapest. http://www.miniszterelnok.hu/prime-minister-viktor-orbans-opening-speech-at-the-2nd-budapest-world-congress-of-families/Accessed: August 18, 2017.

[15] G. Kligman. "Political Demography: The Banning of Abortion in Ceauşescu's Romania." 234-255. In *Conceiving the New World Order: The Global Politics of Reproduction,* ed. Ginsburg, Faye D., Rapp, Rayna. Berkeley: University of California Press, 1995.

[16] A. Miettinen, A. Rotkirch, I. Szalma, A. Donno and M.L. Tanturri, "Increasing Childlessness in Europe: Time Trends and Country Differences." *Väestöliiton Väestöntutkimuslaitoksen työpaperi 2014 Working*

paper 5: 1-55, https://vaestoliitto-fi-bin.directo.fi/@Bin/921a4002358c4b7a03e09df65759e25b/1501911037/application/pdf/3746805/Working%20paper%205_Increasing%20Childlessness%20in%20Europe_1.pdf. Accessed: August 18, 2017.

[17] Hungarian Statistical Office (KSH). *Women Aged 15 and Older and the Number of Live-born Children.* http://www.ksh.hu/nepszamlalas/tablak_teruleti_00. Accessed: August 18, 2017.

[18] J. I. Tóth. "Javaslat a főállású szülő intézményének a bevezetésére" [Proposal for the introduction of a full-time parent institution]. *Valóság* 59(12) (2016): 33-42.

[19] B. Caplan. "The Real Dependency Ratio." *Library of Economics and Liberty.* Permanent Link, January 25, 2008 http://econlog.econlib.org/archives/2008/01/the_real_depend.html Accessed: August 18, 2017.

[20] About Basic Income. *Basic Income Earth Network* (BIEN) https://web.archive.org/web/20080621140909/http://www.basicincome.org/bien/aboutbasicincome.html. Accessed: August 18, 2017.

[21] M. Hermann, T. Pentek, B. Otto. "Design Principles for Industry 4.0 Scenarios." At *System Sciences (HICSS), 49th Hawaii International Conference on System Sciences,* 2016.

Cultural and Ethical Aspects of Social Desirability in Psychological Research

Maria Ancuta Gurza

Introduction

People adopt proactive attitudes in social situations and structure their environment by placing themselves in pleasant situations and in circumstances which ensure their well-being [1]. Some researchers consider that the core of social life is the control of information on one's own identity, activities carried out together with others, and interpersonal relationships [2].

Concepts of social psychology such as impression regulation, impression management, and self-presentation refer the control (intentional or not) of the self-information that people present to others. They do not randomly communicate information on their own person; the situation they are in and the way they relate to it influences how they choose to present information. There is still debate in the literature about the purpose of impression management between those who argue that such behaviour occurs in specific situations and has negative connotations and others who argue that such social behavioural patterns are universal.

These aspects of impression management are especially important when considering evaluating attitudes and behaviours. In this context, parenting, attachment, and couple satisfaction are central elements in life and their assessment is influenced by cultural factors, as well as the context and the way in which the information is requested [3]. Moreover, any intervention in this area can be perceived negatively [4].

Traditional cultural characteristics specific to Romanian society present a challenge to researchers seeking proceeures to access family constructs. Furthermore, it is relevant to take in consideration the respondents' degree of social desirability in order to accurately interpret the data obtained.

Child rearing is a complex activity, defined as a set of normal variations of parental expectations of child control and socialization [5]. Baumrind's model [6] highlights three parenting styles based on two dimensions: parental responsiveness related to parental involvement in relation to children, and parental demandingnesss related to the active role that parents play in promoting compliance with social norms and conventions [7].

Parenting styles, generally speaking, are a function of parental attitudes, beliefs and behaviours, and as some authors [8] argue, they reflect both the parents' goals for their children's socialization and the emotional climate where the parenting practices are implemented [9]. Thus, an authoritative style is often associated with positive effects on the child (e.g. emotional stability, good coping mechanisms, a good level of satisfaction of one's own life) and involves a high level of parental warmth and responsiveness along with a high level of parental control and realistic requirements, having great expectations for their children. These parents provide children with orientation and guidance, but in the context of a warm and loving relationship [7]. Authoritarian style (low responsiveness combined with a high level of unrealistic demands) is characterized by a high expectation of child obedience in the context of a relationship with low affection or a distant parental relationship. This style has been associated with low academic performance and depressive symptoms. In terms of academic performance, studies [10] show that in certain conditions, this style is associated with high academic performance. Parents who practice a permissive parental style, characterized by increased responsiveness and low realistic expectations, give little orientation and guidance to children, and their children are inclined to show low self-control, low self-esteem, and aggression.

The practice of a given parental style was investigated in relation to the personality of the parent, their socio-economic level, cultural affiliation, marital satisfaction, etc. [11]. Studies show that practicing a positive parental style is accompanied by a certain balance in terms of marital satisfaction. Along with this, the insecurity of adult attachment becomes an important aspect of promoting good dyadic relationships. The attachment-related anxiety dimension is defined as adults' feelings, thoughts, and behaviours reflecting their levels of anxiety to the extent to which they feel loved by their partners, and perceive their partners as responsive to their needs. The attachment-related avoidance dimension is defined as adults' thoughts, feelings, and behaviours reflecting their levels of discomfort with closeness and dependency on others in their close relationships [12].

The objectives of the present study were: 1) to identify the difference in social desirability between biological and foster parents; 2) to measure the levels of association between social desirability and the anxiety and avoidance dimensions of attachment, as well as the marital satisfaction in the two groups of parents; 3) to study whether the parental style perceived by the parent as dominant is differently associated with social desirability in the two types of dyads; 4) to investigate whether there are significant differences between the correlation coefficients obtained for the two groups of parents.

The answer to these questions is relevant for understanding the extent to which social cognitions are specific to certain social groups or can be generalized to these groups [13]. This study aims to provide new perspectives into the social desirability of respondents from different social groups,. This is significant for clinical evaluations and interventions in the child-parent dyad, considering the tendency to desirable responses from parents. It is also important to recognise that what counts as a socially-desirable response is closely related to the cultural aspects of research constructs [13] and mirrors how members of a society recognize and adhere to perceived cultural norms [14].

Foster families exist in all cultures as the most effective service for children separated from their biological parents. Compared with biological families, social families are more flexible and more open to those who seek to evaluate aspects of family relationships because they are used to being evaluated by Child Welfare Systems. Kim et al [11] highlight the paradox that these families, although they from a constituent part of the child welfare system in every country, are expected to provide care to children at a standard similar to that provided by the natural family. Social parents go through a selection process, initial and regular training, and have continuous professional supervision in accordance with the law. In other words, these families behave according to professional standards, supported and controlled by the representatives of the protection system.

Foster families are therefore exposed to the intrusion of professionals, and from this point of view are highly different from the biological families whose borders protect their privacy and do not reveal themselves as much to non-family. Working as an interface between the Child Welfare System and the child, the foster family has the subtle task of preserving its unity and its boundaries with the world, like any other family. A placement family therefore has to navigate between the requirements of family integrity and the requirements of healthy child care.

Method

Data collection was carried out through self-reporting within group assessment sessions (four with professional foster carers and four with biological parents). The data was collected in the presence of case managers in the groups of foster parents and of school teachers in the groups of biological parents.

Participants

The participants in the research study included 216 parent-child dyads, of which half (n = 108) were biological families, and half (n = 108) were social families. Five (9%) involved couples of social parents and 98 (91%) single parents, of which 82 (84%) were social mothers and 26 (16%) foster care fathers. These foster families lived in the Timis County (Romania), 41(37,9%) in rural areas and 67 (62%) in the city of Timisoara.

With respect to parental levels of education, 11 mothers and two fathers (N=13; 12%) had completed secondary school (eight school classes), 19 mothers and six fathers (N=25; 23%) had finished vocational high school, 51 mothers and 10 fathers (N=61; 56.5%) had completed academic high school, and six mothers and three fathers (N=9; 8%) were university graduates. The foster parents were aged between 35 and 74 (m=51.86; st.d=8.39).

The general criterion for including these families in the study was to have at least one child in foster care for a minimum of one year. The sub-criteria for inclusion were the following:

- the child was covered by a protection order throughout the period of professional foster care;
- the children's age was between 10 and 17.

The exclusion criteria for the participants in the research were:

- children diagnosed with mental illness, mental retardation, or severe physical disability;
- the period spent by the child in foster was less than one year.

The 108 biological parent-child dyads were recruited in the same period within three schools, one in an urban area and two in rural areas, in two different counties of Romania. The recruitment of families was random, depending only on the willingness of the staff in the schools to be involved. Of the parents, 12 (11%) were single parents (two divorced, four widowed, six mothers who have never married) and 96 (88.8%) had complete families. No couples were included in the data collected. Regarding gender, the group includes four fathers (3,7%) and 104 mothers (96,3%). From an educational point of view, 18 parents (16,6%) finished secondary school, 71 (65,7%) completed high school, and 19 (17,5%) were university graduates. All parents were Christian Orthodox. Their ages ranged between 28 and 59 (m=41.18; st.d.=6.44).

Procedure

The data was collected considering the attachment and parental style for children raised in foster families compared to those raised in biological families.

The foster care families involved in the study were recruited by the Timis DGASPC[22] in response to the researcher's written request. Foster parents

---

[22] The county administrative structure known as the General Social Work and Child Protection Authority.

were invited to the DGASPC headquarters, where between December 2016 and February 2017, groups of 20-25 parents, overseen by the case manager and the researcher, completed self-report questionnaires. The sample consisted of 82 foster mothers (76%) and 26 foster fathers (24%), all registered at DGASPC. As this research was concerned with parent-child dyads, it was carried out in accordance to the ethical research principles of the Ethics Code of Psychology Research and the Society for Research in Child Development (srcd.org). Both biological and foster parents agreed to participate in the research on the basis of informed consent. The children's questionnaires were completed at home and returned to case managers and then to the researcher.

Six classes were recruited from the three schools involved in the research. In the two rural schools, the questionnaires were applied to parents and children (N=68; 62.9%) in the classroom, with the class supervisor and the researcher present. In the urban school, the 40 children and their parents (37%) completed the questionnaires at home and then handed them in to the class supervisor.

Measurements

Although the research used a larger number of measurements to collect data, the following measurements, which are the subject of the present study, are described below.

DECAS Personality Inventory - The "Big-Five" personality profile [15] was employed in order to obtain an adult personality profile [16]. This study used the recorded results only on the social desirability scale, the items of this scale being made and selected taking into consideration the correlations with the two social desirability scales: Marlowe – Crowne Scale and BFQ-2 Scale [15].

The Dyadic Adjustment Scale (DAS) Marriage Satisfaction Questionnaire created by Spanier (1976). The scale contains 32 items divided into 4 subscales, measuring the dyadic consensus (the extent to which the partners agree on important aspects), the cohesion of the couple (the extent to which the partners engage in similar activities), couple fulfilment (the degree of satisfaction in the relationship), and expression of affectivity (sexual and physical affection within the relationship). The internal consistency of the instrument on each subscale is between 0.73 and 0.96, which is a high coefficient. In this research, the Alpha Crombach coefficient of the scale was 0.64 with the following coefficients on each sub-scale: 0.53 for couple fulfilment; 0.67 for the degree of cohesion; 0.40 for consensus; 0.56 for expression of affectivity.

The Parental Authority Questionnaire - PAQ developed by Buri [17] was translated and used in Romanian in an unpublished dissertation, which reported an Alpha Crombach of 0.74 [18]. The assessment forms for parents identify parental practices as reported by parents. When the instrument is used in Romania, the internal consistency of the scale for children is Alpha Cronbach = 0.72, and for the group of parents, Alpha Cronbach = 0.85. The assessment forms for children identify parental practices as reported by children. The scale measured the three parental styles described by Baumrind (1967) [5]: permissive, authoritarian, and authoritative.

The Experiences in Close Relationships-Revised (ECR-R), [19] was used to assess the dimension of adult attachment. This instrument was adapted to the Romanian population [20] and used to investigate the attachment of adults based on two dimensions: anxiety and avoidance [12]. For both the English and the Romanian-adapted versions, the subscales show very high internal consistency (in the English version, Alpha Cronbach for Anxiety = 0.91, and for Avoidance = 0.94, in the Romanian version, Alpha Cronbach for Anxiety = 0.85, and Avoidance = 0.81).

Statistical analysis

This study is a non-experimental one with a standard dyadic design. All statistical analyses were carried out using IBM SPSS Statistics version 20 (SPSS Inc.; Chicago, IL, USA). Descriptive statistics, comprising means, standard deviations, frequencies and percentages were performed to examine socio-demographic characteristics. All data were checked for normality on skewness and kurtosis measures. Information source differences in quantitative variables were tested using the independent sample t test. Pearson's correlations were used to assess associations between variables. The analysis of the difference between two correlation coefficients (to assess whether they are genuinely different from each other) used the q indicator and Δr^2 was used for the effect size. These indicators were obtained using PowerStaTim, Version 1.2.0.0.[20]. All significance tests were conducted on the α=0.05 level and were two-tailed.

Results

Descriptive statistics. The means and standard deviations, for the Parental Authority Questionnaire-Revised (PAQ-R; Buri, 1991), ECR-R, Dyadic Adjustment Scale (Spanier, 1976) and Personality Inventory DECAS [16]are shown in Table 1. The responses to the Parental Authority Questionnaire-Revised range from 1 to 5. In this sample, the Internal consistency for the PAQ-R (a = 0.78) was lower than the alpha of 0.82 reported in Buri [17] . The responses to the Experiences in Close Relationship Scale-Revised (ECR – R; Brennan, Clark, & Shaver, 1998) range from 1 to 7, internal consistency for the ECR-R was a = 0.86. The responses to the Dyadic Adjustment Scale range from 0 to 5. In this sample, the Internal consistency for the DAS was a = 0.65.

The responses to Social desirability range from 0 to 1, Internal consistency for the DECAS Social Desirabiliy is reported as $a = 0.69$[16].

Table 1. Descriptive statistics for parent anxiety and the dimensions of attachment avoidance in romantic relationships, parental styles, and Dyadic Adjustment Scale dimensions within the biological and foster families

		Biological parents		Foster parents	
Variable	*N*	*M*	*SD*	*M*	*SD*
Social desirability (DECAS)	108	15.78	3.04	19.21	2.32
Permissive Style (PAQ-R)	108	32.42	4.83	32.00	4.18
Authoritarian Style (PAQ-R)	108	29.38	8.38	23.80	5.52
Authoritative Style (PAQ-R)	108	40.93	3.48	41.78	3.75
Anxiety subscale (ECR-R)	108	42.65	10.02	38.73	9.79
Avoidance subscale (ECR-R)	108	43.22	16.17	31.55	10.77
Couple fulfilment (DAS)	108	57.92	3.69	59.77	4.32
Dyadic consensus (DAS)	108	34.48	3.34	35.76	3.34
Dyadic cohesion (DAS)	108	18.78	3.34	19.87	3.50
Expression of affection (DAS)	108	14.70	1.35	15.25	1.52

Major findings

In order to investigate the differences between biological and foster parents, we applied the t test for independent samples. The results obtained are illustrated in Table 2.

A significant difference can be seen in social desirability between biological and foster parents, as foster parents obtained significantly higher scores than biological parents [t(200.04)=-9.280]. Furthermore, the magnitude of this correlation is very strong (Cohen's d=1.31).

Table 2. The **significance of parent social desirability differences in the two types of families (biological/social)**

	Family	N	Mean	Std. Deviation	t	df	p
Social desirability	Biological	108	15.78	3.04	-9.280	200.04	**.000**
	foster	108	19.21	2.32			

In order to identify the strength of association between social desirability, parental styles, the anxiety and avoidance of adult attachment in close relationships, as well as marital satisfaction, we used the Pearson correlation coefficient; Table 3 comparatively illustrates the significant associations found in the two types of families.

The social desirability of the biological parents in this study is negatively associated with the permissive style [r (108) = -0.269, p <0.01] and with the authoritarian style, [r (108) = -0.293, p <0. 01]. A high level of social desirability is also associated with a high level of avoidance in adult attachment, [r (108) = -0.344, p <0.01]. A positive correlation was found between social desirability and the dyadic cohesion of biological parents [r (108) = 0.298, p <0.05]. These findings point to the fact that the biological parents who obtained high scores on the social desirability dimension tend to appreciate less permissive and less authoritarian parental styles. A high level of social desirability is also associated with a high avoidance of attachment in close relationships and high dyadic cohesion.

The social desirability of the foster parents is negatively associated with the authoritarian style, [r (108) = -0.243, p <0.05], and with the avoidance of adult attachment in romantic relationships, [r (108) = -0.249, p <0.05]. In other words, the higher the social desirability level, the more they appreciate less authoritarian parenting styles and manifest higher avoidance in romantic relationships. This tendency to offer social desirable responses is accompanied by the tendency towards a rather authoritative parental style, [r (108) = 0.246, p <0.05], while couple fulfillment is higher. [r (108) = 0.311, p <.01].

As shown in Table 3, the two groups of parents are similar in their associations between social desirability and authoritarian style and between authoritarian style () and the avoidance dimension of adult attachment. There are some differences between the two groups of parents on the level of the other variables associated with social desirability.

Table 3. The signification of the correlations between social desirability and parental styles, attachment dimensions, couple satisfaction

	Social desirability					
	Biological family N=108		Foster family N=108		The effect sizes	
	r	p	r	p	q	Δr^2
Permissive style	**-.26****	.00	.15	.11	.12	.04 (low)
Authoritarian style	**-.29****	.00	**-.24***	.01	.05	.02 (low)
Authoritative style	.01	.86	**.24***	.01	.23	.06 (moderate)

Anxious dimension of attachment	-.15	.11	-.13	.17		
Avoidance dimension of attachment	**-.34****	.00	**-.24****	.00	.10	.05 (low)
Couple fulfilment	.04	.65	**.33****	.00	.30	.11(moderate)
Dyadic consensus	.14	.14	-.08	.37		
Dyadic cohesion	**.29****	.00	.06	.50	.24	.08 (moderate)
Expression of affection	.13	.15	.19*	.04		

$p<.05^{*}$, $p<.01^{**}$

The question we sought to answer in this study was whether social desirability is a predictor of the positive appreciation of marital relationships and the parent-child relationships for biological parentsrather than for foster parents. So, we calculated a base indicator q and the effect size noted as Δr^2, which represent the magnitude of the difference between the two r^2 coefficients obtained [15]. Table 3 shows that the effect size of the differences between the correlation coefficients are low and moderate.

Thereffore, regardless of their type of family, parents with a high level of social desirability have a tendency to appreciate less authoritarian and less permissive parental styles. Also, parents with a high level of attachment avoidance in romantic relationships tend to exhibit a high level of desirability.

The magnitude of the difference between the correlation coefficients was average on the level of the authoritative style and the two dimensions of marital satisfaction: dyadic cohesion and couple fulfilment. Thus, social parents who manifest a higher level of social desirability tend to appreciate more authoritative parental styles and show higher levels of couple fulfilment. Biological parents who obtained higher scores on the social desirability scale tend to have a higher appreciation of dyadic cohesion.

Discussion

The main purpose of this study was to identify the differences in the social desirability between biological and foster parents in order to identify the extent to which that dimension can be considered a good predictor of positive appreciation of the marital relationship and parent - child relationship.

We considered that a positive appreciation of the couple's relationship is manifested by high scores in the dimensions of marital satisfaction and a low level of anxiety and attachment avoidance in romantic relationships. The positive appreciation of the parent-child relationship was operationalized by the extent to which the parent considered that he/she practiced a parental responsiveness (permissive and authoritative) style rather than a parental demandingness (authoritarian) style [5].

We expected that foster parents would obtain lower scores on the social desirability scale because of their frequent contact with the Child Welfare System compared to the biological parents who are only ocassionally asked by the school to evaluate their family relationships.

The wide-ranging framework of impression management underlines the fact that, in the self-assessments of attitudes or behaviours, individuals can claim desirable attributes not only directly, by presenting information on their own qualities, but also indirectly, by presenting information on the qualities of others. Our findings demonstrate these theoretical observations in the correlations shown between the two dimensions of Dyadic Adjustment Scale (dyadic cohesion and couple fulfilment) and social desirability.

Social desirability becomes important for discussing the precision of attitude measurement. Attitudes become acute when targeting differently validated cultural and social behaviours. In these two groups of parents, both the assessment constructs (attitudinal/emotional), the way in which the study was carried out (group assessment) and the (institutional) framework in which it took place, and given the specific cultural context of the research, as well as the limitations of its proper ethical constraints, these findings need to be interpreted with an appropriate level of caution.

However, such considerations do require us to discuss concepts such as self-presentation, desirable identity, and social desirability. Thus, regardless of the type of family they belong to, parents who appreciate their parental style as authoritarian tend to provide socially-desirable responses. They also show a high level of attachment avoidance in romantic relationships, which, according to attachment theory, shows a lack of affective involvement in their relationship with the child[21].

Some authors [9] consider that for good child development in a foster family, it is necessary to meet the conditions considered in the stages of recruitment, the monitoring and evaluation of parents: the desire of the family members to care for a child and their motivation; good parental competence including quality of child care, promotion of children's independence, non-punitive discipline and the quality of specific treatments; emotional security of parents.

These dimensions are periodically evaluated by case managers and psychologists who monitor the foster family. This way, the support provided by child welfare workers to foster parents in terms of how to interact with the child, but also aspects of the couple relationship, lead to a higher likelihood that they will provide answers presenting an enhanced level of social desirability in the context of a group evaluation. In turn, biological parents are less well-informed about these aspects of family relationships, and this may be an explanation of their lower desirability in responding. This observation is also supported by the differences in the relationships between social desirability and the dimensions of marital satisfaction appreciated by the two types of parents.

The differences between the two parental groups regarding couple fulfilment and dyadic cohesion support the idea that desirable answers were selected by the foster parents, in the sense that their social desirability is associated with the emotional stability dimension, an aspect directly targeted during the evaluation for child placement in a foster family. Whereas in the group of biological parents, the tendency to desirability responses is associated with the answers given on the level of dyadic cohesion, a more pragmatic aspect of the couple relationship, referring to commitments and involvement in collective activities.

The results of this study are important both for identifying cultural influences on family life and for emphasizing aspects of professional ethics in relation to the three actors: children, foster parents and the Child Welfare System. Although research on foster parents in Romania is still at the beginning, this study highlights important aspects and limitations which must be considered for the accurate interpretation of findings.

Some limitations of the study need to be considered. Firstly, the procedure of recruitment based on the goodwill of institutional staff requires caution caution when generalizing the results. Also, as a comparative study, the results cannot be used to infer causal relationships between the variables. The results of this study emphasize once again the importance of discussing social desirability in the cultural context of both family and institutional relationships, and it is possible to assert that professional foster parents incline to manifest a certain degree of institutionalised behaviour. For a clearer image of the associations of these constructs, as well as for a study of possible causal conclusions, we propose the elaboration of more complex methodological studies involving this particular social group. The aspects discussed here highlight important ethical and cultural factors which could be useful in clinical intervention and practice involving parent-child dyads regardless of whether they are biological or formed through a fostering relationship.

References

[1] S. Boncu. *Psihologie si Societate* [Psychology and Society]. Iasi: Editura EEOTA, 1999.

[2] E. Goffman. *Behavior in Public Places.* New York: The Free Press, 1963.

[3] M. Kemmelmeier. "Cultural differences in survey responding: Issues and insights in the study of response biases." *International Journal of Psychology* 51(6) (2016): 439–444, doi: 10.1002/ijop.12386

[4] M. Roth, C. Voicu, A. Dávid-Kacsó, I. Antal, A. Muntean, S. Bumbulu, C. Baciu. "Asking for parental consent in research on exposure of children to violence." *Revista de cercetare si interventie sociala* 42 (2013): 85-100.
[5] Baumrind, D. "Child care practices anteceding three patterns of preschool Behavior." *Genetic Psychology Monographs* 75(1) (1967): 43-88.
[6] C. H. Hart, L.D Newell, S. G. Olsen, John O. "Parenting skills and social-communicative competence in childhood." 753-797. In *Handbook of communication and social interaction skills*; ed. Burleson, B. R. Mahwah, NJ: Lawrence Erlbaum Associates Publishers, 2003.
[7] Akhter, N., Hanif, R., Tariq, N., & Atta, M. "Parenting styles as predictors of externalizing and internalizing behavior problems among children." *Pakistan Journal of Psychological Research* 26(1) (2011): 23-41. Available at https://search.proquest.com/docview/920830272?accountid=136549.
[8] N. Darling, L.Steinberg. "Parenting style as context: An integrative model."*Psychol Bull* 113 (1993): 487–496.
[9] T. G.Power, E. F. C. Sleddens, J. Berge, , L. Connell, B. Govig, E. Hennessy, S. M St George. "Contemporary research on parenting: Conceptual, methodological, and translational issues." *Childhood Obesity* 9: 87-94. doi: 10.1089/chi.2013.0038.
[10] Lola, B., Shrinidhi, I. "Parenting styles: the impact on student achievement." *Marriage Fam. Rev.* 43 (2008): 14–38.
[11] J. Kim , J. L. Shawna, C. A. Taylor , and N. Guterman. "Dyadic Profiles of Parental Disciplinary Behavior and Links With Parenting Context." *Child Maltreatment* 19(2) (2014): 79-91, doi: 10.1177/1077559514532009
[12] K. A.Brennan, L. C. Clark, & , P. R. Shaver. " Self-report measurement in adult attachment: an integrative overview". In *Attachment theory and close relationships*, ed. J. A. Simpson, & W. S. Rholes. New York: Guilford Press, pp. 46–76, 1998.
[13] M.H. Bornstein, "Children's parents." 55–132. In *Ecological settings and processes in developmental systems*, ed. Bornstein MH, Leventhal T, vol. 4. Handbook of child psychology and developmental science. Hoboken, NJ: Wiley, 2015.
[14] P.B. Malham, G.Saucier. "The conceptual link between social desirability and cultural normativity." *International Journal of Psychology* 51(6) (2016): 474-480, doi:10.1002/ijop.12261
[15] F.A. Sava. *Analiza datelor in cercetarea psihologica* [Data analysis in psychological research]. Cluj-Napoca: Editura ASCR, 2011.
[16] F.A. Sava , R.I. Popa. "Personality types based on the Big Five Model. A cluster analysis over the Romanian population." *Cognition, Brain, Behavior. An Interdisciplinary Journal* 15(3) (2011): 359-384. ASCR Publishing House.

[17] Buri, J.R.. "Parental authority questionnaire." *Journal of Personality Assessment* 57 (1991): 110-119. doi:10.1207/s15327752jpa5701_13.

[18] A.M. Diagostin. "Stil parental si depresia la adolescenti" [Parental style and adolescent depression], PhD dissertation, Available on http: //dspace. biblioteca. um.edu.mx/ xmlui/bitstream /handle /20.500.11972/408/Tesis%20Anca-Mariana %20 Dragostin. pdf?sequence =1&isAllowed=y.

[19] R.Fraley , N. Waller & K.Brennan. "An item response theory analysis of self-report measures of adult attachment." *Journal of personality and social psychology* 78(2) (2000): 350–365.

[21]Negrei, N., Sava, F.A., "Adaptation of the adult attachment styles inventory for the Romanian population". *Social Psychology* 18 (2006): 155-169.

[21] H. Masud, , R. Thurasamy, M. S. Ahmad. "Parenting styles and academic achievement of young adolescents: A systematic literature review." *Quality and Quantity* 49(6) (2015): 2411-2433, available on http://dx.doi.org/10.1007/s11135-014-0120-x

Author Index

FSC
www.fsc.org
MIX
Papier | Fördert gute Waldnutzung
FSC® C083411

Kolibri 360 GmbH
Ferdinand-Jühlke-Straße 7
99095 Erfurt, Deutschland
produktsicherheit@kolibri360.de